How It All Works

IVY PRESS

How It All Works

All Scientific Laws and Phenomena
ILLUSTRATED & DEMONSTRATED

Adam Dant & Brian Clegg

CONTENTS

Science is the study of how the constituents of the universe work. This could be an abstract pursuit of knowledge for its own sake – and that is a worthwhile activity in its own right. But since the development of applied scientific principles, notably in optics, it has also gradually become a practical study that enables us not only to understand how things work, but how to make use of that understanding in the technology that supports our everyday lives.

Laws and phenomena

In Adam Dant's magnificent and ever-delightful illustrations we can see a whole host of ways in which scientific laws and phenomena crop up in everything we do. The distinction between laws and phenomena is a subtle one. A phenomenon is something that happens or exists in the universe. It can be anything from an object – a star, for example – to a mechanism by which something takes place, for example, the way that a fluid flows from place to place, or life itself. Scientific laws, though, are more a way of describing structures that link the different phenomena.

Unlike the human-devised legal system, the universe doesn't have a rulebook where can we look up precisely what to expect in a particular situation. Instead, a scientific law is an attempt to reflect a repeated pattern in nature. The great American physicist Richard Feynman remarked that: 'There is ... a rhythm and a pattern between the phenomena of nature which is not apparent to the eye, but only to the eye of analysis; and it is these patterns which we call Physical Laws.'

What we describe as a law is often a mathematical description of what is expected to happen. Some are the pragmatic result of observation. So, for example, we will meet Kleiber's law, which relates the weight of an animal to the amount of energy it consumes. There is no theory here – such laws are a way of stating that 'we have observed a lot of these things, and they usually behave like this'. Other laws, notably physical laws, are driven by theory and work consistently within particular limits. So, for example, Newton's laws of motion describe exactly how an object will move as long as it is not moving too quickly.

How the book works

Each of the illustrations in the book covers a particular level in a cosmic zoom that starts with a kitchen and moves outward, step by step through the house, the garden, a science museum, a hospital, a town square, a street, the countryside, the coastline, a continent, planet Earth, the solar system and the universe – though thanks to Adam's imagination, some of the locations may be unexpected. Within each illustration you will be able to find 46 different laws and phenomena behind the rich play of activities and objects featured – the laws are marked 'L' and phenomena with 'P'.

Each item covered is then pulled out as a detail to describe what is being illustrated. Inevitably, there is a limit to the amount of information that can be included in such a short piece of text. As a result, it is sometimes necessary to omit details in what can be a quite complex concept. In most cases, you can find out more by searching for the topic on the internet – though it's fair to say that in some aspects of, say, quantum physics, even the scientists can struggle to get their head around what's happening. Similarly, it's impossible to be comprehensive because of the sheer scale of the ever-changing world of science, but the topics are representative of the scientific landscape at the time of publication.

What these illustrations and their short descriptions will show is the way that in everything we do, in everything we experience, we are witnessing and taking part in scientific phenomena, guided and linked by scientific laws. Science is not just something we do at school or that professionals undertake in laboratories, it is at the heart of how everything works. Samuel Johnson famously commented that 'when a man is tired of London, he is tired of life'. In this book we see how anyone who says that they have no interest in science is also saying they have no interest in life, the universe and everything.

Key figures

At the back of the book, you will find 13 'game changers' – key figures who have had a significant impact on our scientific understanding of how it all works. There is one of these for each section (you will find each of the key figures cropping up in one of the illustrations, where his or her discovery features). Any attempt to nominate a scientific hall of fame is fraught with difficulties. I have tried in selecting these individuals to go for a mix of well-known and more obscure names, but in each case the person has been responsible for giving us a step forward in the understanding of how our universe works.

Because these scientists have given us insights into fundamental laws and phenomena, many of them lived before the twentieth century. There have been vast numbers of scientific breakthroughs since 1900, yet outside of a few areas such as quantum physics and chaos theory, the basics were

already established. This does mean that we feature only two women among our 13 key figures. If we were looking at leading scientists in the last 50 years, say, the proportion would be very different. Even before the twentieth century there were many female contributors to scientific knowledge, but because of cultural restrictions in those days, the proportion of women in science was far lower. Thankfully, things have now changed.

The beauty of science

It's possible to enjoy *How It All Works* on two complementary levels. We have here a wonderful collection of pictures by Adam Dant, demonstrating his mastery of large-scale illustrations with the richness that won him the Jerwood Drawing Prize. However, Adam's pictures are more than works of art. When we look deeper, each illustration shows us in a multi-layered fashion how the many aspects of science and technology come together to establish our relationship with the universe we inhabit.

Famously, the nineteenth-century English poet John Keats complained that Newton was guilty of 'unweaving the rainbow', reducing the beauty of nature to mere mathematics. However, as Adam's illustrations show, there is no dividing line between science and beauty. With a scientific understanding we can both appreciate the splendour of nature and get a better feeling for how it all works. And that isn't a bad thing.

Brian Clegg

The Kitchen

The Kitchen

Charles' law [L]
At constant pressure, the volume of a gas is proportional to its temperature. Bubbles of air grow in volume when heated to give the cake its fluffy texture.

Gay-Lussac's law [L]
The temperature of a gas varies with pressure. In the fridge, the refrigerant (the chemical that keeps it cold) is compressed then expanded to transfer heat from the interior to the radiator on the back of the fridge.

Faraday's law of induction [L]
The voltage induced in a circuit is dependent on the rate at which the strength of the magnetic field crossing it changes. The hob induces a current in the pan to heat it.

Joule's first law [L]
The heat generated by an electrical element is proportional to its resistance and the square of the current. This is the basis for how the electric toaster works.

First law of thermodynamics [L]
The change in the internal energy of a system is the heat supplied, less the work done by the system on its surroundings. Energy in the pan is increased by heat from the hob.

Second law of thermodynamics [L]
In a closed system, entropy remains the same or increases. Entropy is the number of ways in which the parts of a system can be arranged: the broken china has many more ways to be arranged, increasing entropy.

Zeroth law of thermodynamics [L]
If two systems are in thermal equilibrium with a third system, they are in thermal equilibrium with each other. This allows a thermometer to compare temperatures.

Capillary action [P]
Liquid flows into narrow gaps against the force of gravity because it is attracted to the sides of the surrounding material. Capillary action in the kitchen towel helps it absorb the spilled liquid.

Acid–base reaction [P]
Acids and bases react by transferring a pair of electrons from one compound to another. Vinegar (acetic acid) and baking powder (containing the base sodium bicarbonate) react, giving off carbon dioxide.

Caustic curve [P]
Reflections and refractions at the surface of a liquid can produce bright curves of light. This is often seen in a cup or bowl of liquid.

Adiabatic expansion [P]
A reaction where energy is output as work rather than heat. The water in the kernels turns to steam, fluffing up the grains and reducing the temperature as the gas expands.

Cohesion [P]
When molecules are attracted to other molecules of the same substance they pull together. On a repulsive surface like the wax carton, cohesion produces near-spherical drops.

Borborygmus [P]
The rumbling noise as stomach contents are pushed through the small intestine by muscle contractions.

Conduction [P]
Heat flows from place to place as fast-moving molecules in a substance bump into other molecules and start them moving. The oven glove prevents the heat from the metal tray being conducted to the man's hand.

Convection [P]

The higher speed of warmer molecules reduces the density of a fluid, so warm air moves upwards. Smoke particles from the grill are carried with warm air upwards to the smoke detector.

Covalent bonding [P]

Many compounds are held together by covalent bonds, where outer electrons are shared between atoms. Sucrose, the compound in table sugar, has many such bonds.

Dissolution [P]

When a solid dissolves, bonds between the solid's molecules are broken. The hot water in the cup of tea dissolves the sugar because the water molecules are better at attracting sugar molecules than the sugar itself.

Eddy currents [P]

Electrical currents often flow around circuits, but they can also be induced as loops within a conductor. The induction hob produces eddy currents in the base of the pan, heating it.

Electroluminescence [P]

Some materials, particularly semiconductors, give off light when electricity is passed through them. The LEDs in the light-up clock are electroluminescent.

Emulsion [P]

Some liquids that don't normally mix can form a homogeneous fluid. In homogenised milk, liquid butterfat is broken down into very small globules to form an emulsion with water.

Enzymatic hydrolysis [P]

Enzymes are catalysts, encouraging chemical reactions. When the bread is chewed, the enzyme amylase in saliva breaks down starch into sugars and adds water, helping digestion.

Exponential growth [P]

Growth that increases by the same factor every time period, for example, doubling every hour. Bacteria on rotting fruit grow rapidly by doubling.

Fermentation [P]

Fermentation occurs when microorganisms such as bacteria or yeast break down carbohydrates to produce alcohol, lactic acid and carbon dioxide. Beer is produced when yeast breaks down starch in malted barley.

Ferromagnetism [P]

Permanent magnets, working through ferromagnetism, have tiny crystals in their structure aligned so their magnetic fields add together. Fridge magnets incorporate small ferromagnets that enable them to attach to the metal fridge.

Fluorescence [P]

A substance that gives off light after being stimulated by another light source, often of a higher energy. In the white LED light, blue light inside stimulates the fluorescent coating to give off white light.

Friction [P]

Interaction between surfaces resisting movement. The rough surfaces of the man's rubber-soled shoes catch on tiny bumps in the flooring, making it safer to carry the tray without slipping.

Ionic bonding [P]

Bonds between ions – atoms with either excess or missing electrons – giving them an attracting electrical charge. The salt being poured into the tea has ionic bonds between sodium and chlorine ions.

Iridescence [P]

Colours generated in thin transparent layers where reflections from different surfaces interfere. The colours in the soap bubbles in the kitchen sink are iridescent.

Latent heat [P]

When a substance boils, heat being put into it no longer increases the temperature but vaporises the liquid. The boiling water in the pan on the hob stays at 100°C (212°F).

Leidenfrost effect [P]

When a liquid is on a surface hotter than its boiling point, drops float on a layer of vapour. On the hob, droplets dance across the surface.

Maillard reaction [P]
In many foodstuffs, heat produces a reaction between sugars and amino acids that browns the food and produces savoury aromas. Roast meat undergoes the reaction when heated in the oven.

Non-Newtonian fluid [P]
A non-Newtonian fluid becomes thicker or thinner under pressure. Ketchup becomes runnier due to the pressure wave that is generated by tapping the bottle.

Meniscus [P]
Because many fluid molecules are more strongly attracted to the walls of the container than themselves, they climb the container. The surface of the water in the glass is concave due to the meniscus.

Ostwald ripening [P]
When small crystals melt and then reform as larger crystals. Melted ice cream that has been refrozen undergoes Ostwald ripening, degrading its texture.

Metallic bonding [P]
Inside metals, the bonding of the lattice of atoms allows electrons to move freely. The free electrons carry electrical current from the socket to the radio.

Parabolic trajectory [P]
When an object is projected upwards and forwards under the force of gravity it traces the curve of a parabola. The thrown cherry follows a parabolic trajectory.

Moiré patterns [P]
When light passes through two grids that are not quite identical, an interference pattern of dark and light forms. The lattice on two sides of the lampshade creates these moiré patterns.

Peristalsis [P]
A wave of muscular contractions to propel food through the body's gastrointestinal system. Peristalsis carries the chewed bread down to the girl's stomach.

Quantum leap [P]
Electrons around an atom change energy levels in jumps called quantum leaps, corresponding to the energy of a photon (particle of light) given off or received. The lamp produces light when electrons in the bulb undergo quantum leaps.

Starch gelatinisation [P]
Heat and water content break down covalent bonds in starch molecules, giving a smoother texture. In the baked potatoes this process makes the potato flesh fluffier.

Radiation [P]
Heat can be transferred by electromagnetic radiation, most effectively in the infra-red. The heater transfers heat to its surroundings by radiation.

Strong interaction [P]
The force that holds together quarks, the components of protons and neutrons, and leaks out to hold atomic nuclei together. All matter – from the birds to the air they fly in – is dependent on the strong interaction to remain stable.

Rayleigh–Bénard convection [P]
When a liquid is heated from below, heat flows upwards in circulating cells, producing a pattern on the surface of the soup.

Waveguide [P]
A structure that guides waves with minimal loss of energy. In the microwave a rectangular metal tube guides the microwaves from generator to cooking chamber.

Sporulation [P]
Reproduction by the production of spores, common in fungi, plants, algae and protozoa. Mould grows on rotting fruit from fungus spores spread in the air.

Weakness of gravity [P]
Gravity is the weakest force of nature, a trillion trillion trillion times weaker than that of electromagnetism. The electromagnetic force of the fridge magnet is stronger than the gravitational attraction of the whole Earth.

The House

The House

Ampère's law [L]
Describes the attractive force between two current-carrying wires. The washing machine uses a solenoid valve, based on this force, to control water flowing into the machine.

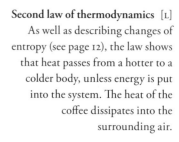

Second law of thermodynamics [L]
As well as describing changes of entropy (see page 12), the law shows that heat passes from a hotter to a colder body, unless energy is put into the system. The heat of the coffee dissipates into the surrounding air.

Lorentz force law [L]
Shows how an electrical current acts as a magnet. The basis of electrical motors, such as the motor in the washing machine.

Acid [P]
An acid acts by accepting an electron pair from another atom or donates an excess proton (hydrogen ion). Acids such as vinegar or formic acid remove limescale (calcium carbonate) in such a reaction.

Pascal's principle [L]
Relates how applying pressure to one part of a contained fluid transmits it to the rest. The principle behind hydraulic lifts also explains how squeezing the shampoo bottle causes liquid to squirt out.

Alkali [P]
An alkali, which is a water-soluble base, acts by donating an electron pair to another atom or receiving a proton that attaches, for example, to a hydroxide ion to form water. The cleaning product contains an alkali, which reacts with grease to produce a soap-like material.

Capacitive screen [P]
A touchscreen where the finger's position is detected by changing the capacitance (ability to store electrical charge) of components below the surface of the screen. The tablet's touchscreen is capacitive.

Chimney effect [P]
The higher temperature at the bottom of the chimney causes the air to rise up, pulling fumes with it.

Catalysis [P]
A catalyst increases the rate of a chemical reaction without being consumed. In the biological detergent, naturally occurring catalysts called enzymes help to break down stains.

Condensation [P]
If a surface is cool enough it will cause water vapour in the air to condense out when it comes into contact. The window is cooled by contact with the outside, causing water droplets to condense out of the steamy bathroom air.

Centre of gravity [P]
Objects act under gravity as if all their mass were concentrated at this point. The boy's centre of gravity is over to one side, encouraging him to topple over.

Cosmic rays [P]
The Earth is constantly bombarded with energetic particles from outer space called cosmic rays. When one of these charged particles passes through the memory or processor of a computer it can change a value and cause the computer to crash.

Cheerios effect [P]
Named after the breakfast cereal, the tendency of small floating objects to attract each other, as surface tension means that some parts of the surface are higher than others. As a result, the ducks float together.

Dissolution [P]
The molecules in the bath salts have relatively weak bonds, easily broken by the hot water.
See also page 14.

Electromagnetism [P]
Atoms are almost entirely empty space. It is the electromagnetic repulsion between charged particles in atoms that stops objects passing through each other. The dog does not fall through the ground due to electromagnetism.

Feedback effect [P]
When information about the state of a system is used to modify that state. As the temperature reaches the value set on the central heating thermostat the boiler cuts out until the temperature drops.

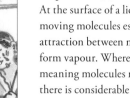

Erythema [P]
Reddening of skin when blood flow is increased due to inflammation or injury. The hot water in the mop bucket reddens the person's hand.

Feigenbaum number [P]
A constant of nature: some chaotic systems repeatedly increase the number of possible modes, with the time between such splits dropping by this factor. As water pressure is increased, the rhythm of the drips from the tap changes in response to this constant.

Evaporation [P]
At the surface of a liquid, fast-moving molecules escape from the attraction between molecules to form vapour. Where water is hot, meaning molecules move faster, there is considerable evaporation.

Insulation [P]
Using materials that are bad conductors of heat to prevent heat loss. The air gaps between glass sheets in the double-glazed windows act as insulators.

Evaporative cooling [P]
When a liquid evaporates from a surface, energy is extracted, lowering the temperature of the surface. The damp person shivers because of evaporative cooling.

Kaye effect [P]
Liquids that split cleanly under sideways stress briefly produce an upward jet of liquid as the liquid is poured on to a surface. Shampoo exhibits the Kaye effect.

Liquid crystals [P]

A substance part-way between a liquid and a crystal. Some change polarisation when a voltage is applied, enabling them to control how much light passes through. The tablet's display is based on liquid crystal technology.

Maxwell's colour triangle [P]

Shows how mixtures of different intensities of red, green and blue produce all possible colours. A colour screen like that on the smartphone produces colours using pixels with a combination of red, green and blue.

Mechanical advantage [P]

The amplification of force produced by a machine such as a lever. Tweezers are class 3 levers with a mechanical advantage of less than 1, because the force applied is closer to the fulcrum than that of the resultant movement.

Opacity [P]

Light is prevented from passing through matter when it is absorbed by the substance and not re-emitted to travel on in the same direction. The towel hides what's beneath because the cloth is opaque.

Partial reflection [P]

Some of the photons hitting a transparent substance can reflect while other pass through: this is a quantum effect. Although most photons pass through a window, a few are reflected back: we see the reflection because it is relatively dark on the other side.

Photoelectric effect [P]

Some materials, notably semiconductors, produce an electrical current when exposed to light. The solar panels generate electricity from sunlight.

Principle of least action [P]

The free movement of objects in flight minimise a property called action, which is dependent on the difference between kinetic energy (the energy of movement) and potential energy (stored energy). The path taken by the drops from the showerhead obeys the principle.

QED reflection [P]

Quantum electrodynamics, the theory of the interaction of light and matter, predicts that mirrors with pieces missing will reflect at unusual angles. The CDs produce rainbow reflections at odd angles as the pits in the discs that store the music block parts of the reflection.

Quantum tunnelling [P]
Because their locations are probabilistic, quantum particles such as electrons and photons can appear the other side of a barrier without passing through it. The memory stick uses quantum tunnelling in order to store information in flash memory.

Siphon [P]
Makes use of gravity and changes in pressure to carry a liquid from a higher to a lower location over an intermediate high point. This is happening from the toilet bowl: some toilets also use siphoning action to take water over a U-bend.

Radioactivity [P]
When atomic nuclei are too big, they decay, breaking into smaller parts and emitting particles in the process known as radioactivity. The smoke detector uses this ionising effect of radioactive particles to detect smoke.

Specular reflection [P]
Classical reflection of waves, as light does from a mirror, reflecting at the same angle as the incoming light. The bird can see its specular reflection in the window.

Reverberation [P]
When sound doesn't decay immediately but is reflected off surfaces, keeping the sound audible longer. The singer's voice has reverberation thanks to the hard, tiled surfaces.

Strong interaction [P]
The strong interaction holds together the quarks that make up protons and neutrons in the atomic nucleus: most of the mass in matter is from the energy of the strong interaction. The majority of the man's weight is from the strong interaction in his atoms.

Shower curtain effect [P]
A running shower causes a shower curtain to billow inwards: this may be because the flow of water reduces air pressure. The man in the shower has the curtain move towards him.

Surface tension [P]
Molecules of a liquid such as water are attracted to each other: at the surface there is more attraction into the body of the liquid, causing the surface area to be minimized. This forms the water drops from the tap.

Temperature [P]

A measure of the thermal energy in matter – the faster the atoms or molecules move and vibrate, the higher the temperature. In the hot bath water, the molecules are fast moving.

Translucence [P]

A translucent material lets light through, but scatters the photons so that no clear image of the other side is visible. The frosted glass in the door is translucent.

Transparency [P]

A transparent material allows light through without significant scattering as relatively few photons are absorbed and re-emitted in new directions by the atoms. Glass is transparent for this reason.

Triboelectric effect [P]

The production of electricity by rubbing a material where electrons are easily removed this way from the outside of atoms. The artificial fibres in the man's shirt are triboelectric, which causes small electric shocks as it is taken off.

Venturi effect [P]

When a fluid passes through a constriction, pressure drops and velocity increases. In the perfume atomiser, the Venturi effect produces a fine spray of droplets.

Virtual image [P]

The image in a mirror is not tangible, but appears to be the same distance behind the mirror as the object reflected is away from the mirror's surface. The reflection of the man shaving appears to be behind the mirror.

Vortex [P]

A fluid flow that rotates around a central axis. The water running down the plughole forms a vortex (but it's a myth that the direction is controlled by location on the Earth).

Wheel and axle [P]

The wheel and axle is a machine that amplifies force applied to the outside of the wheel. When the toilet paper is pulled it exerts enough force on the axle to turn the roll, but the lower force on the paper does not break the paper.

The Garden

BEER

The Garden

Avogadro's law [L]
This law states that equal volumes of gas at the same temperature and pressure contain the same number of molecules. The helium molecules in the balloons are lighter than the those in the gasses in air, so with equivalent numbers of molecules, the balloons float.

Conservation of momentum [L]
Momentum – the mass of an object times its velocity – is subject to a conservation law. The total amount of momentum in a system is conserved: when the girl on the swing hits someone, she loses momentum and the boy that she hits gains momentum.

Brewster's law [L]
According to the law, when light hits a substance at a particular angle, only polarised light reflects – this is light where the waves move side to side in a fixed direction as they travel. The light reflected from the pool is polarised.

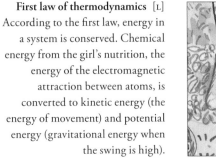

First law of thermodynamics [L]
According to the first law, energy in a system is conserved. Chemical energy from the girl's nutrition, the energy of the electromagnetic attraction between atoms, is converted to kinetic energy (the energy of movement) and potential energy (gravitational energy when the swing is high).

Bunsen–Roscoe law [L]
This reciprocity law says that the reaction of a light-sensitive material is dependent on the intensity and duration of exposure to light. In the twilight, the eyes struggle to pick up enough light to make out detail.

Gay-Lussac's law [L]
States that for a fixed volume, the pressure of a gas varies with temperature. When the firework ignites, the confined, rapidly heated gas produces a high pressure, blasting apart the container.

Hooke's law [L]

If a spring is not stretched too far, the law states that the force needed to extend it increases with the distance the spring is extended. The further the catapult is pulled back, the harder the boy has to pull on it.

Allotropes [P]

Allotropes are variants of a chemical element with different structures and properties. The barbecue charcoal is one allotrope of carbon, as are diamond and graphite.

Le Châtelier's principle [L]

Sometimes called the equilibrium law: in effect, when a change is made to a system it causes a reaction that acts to reduce the change. Shaking the bottle increases the pressure from the gas inside: when the bottle is opened, liquid squirts out, reducing the pressure.

Anisotropy [P]

A material with properties that are dependent on direction is anisotropic. It is easier for the man to chop the wood along the grain because it is anisotropic.

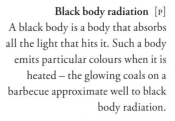

Newton's third law [L]

The law states that every action has an equal and opposite reaction. The firework rocket blasts propellant out of the back: as a reaction the rocket is pushed into the sky.

Black body radiation [P]

A black body is a body that absorbs all the light that hits it. Such a body emits particular colours when it is heated – the glowing coals on a barbecue approximate well to black body radiation.

Second law of thermodynamics [L]

According to the second law, heat moves from a hotter to a colder body – in this case from the hot sausage to the man's hands.

Cavitation [P]

When bubbles in a liquid collapse they can produce a strong shock wave. The fern expels spores at high speed using this cavitation effect.

Chemiluminescence [P]

The emission of light by a chemical reaction. The glow sticks work through chemiluminescence when an internal container is snapped, mixing chemicals to start a reaction.

Fluorescence [P]

When an object absorbs light, then re-emits it in a different frequency (colour), it is fluorescent. Some flowers appear brighter than expected at dusk because they absorb ultraviolet light and fluoresce in the visible range.

Cocktail party effect [P]

The brain enables us to pick out a conversation in a crowded area where many are talking. Thanks to the effect, despite the noisy party, the individuals can carry on separate conversations.

Green flash [P]

As the Sun sets over the horizon or clouds a brief green flash can occur, as an optical effect called refraction. This momentarily splits out some of the colours of the light from the top of the Sun.

Echolocation [P]

The use of reflected sound to detect the position of objects. To catch insects in low-light conditions, the bats give off high-pitched sounds and use the reflections to detect the location of the insects.

Guttation [P]

Some plants exude water drops at night when they have absorbed too much water through their roots. Both grass and strawberry plants exhibit such guttation.

Fibonacci number [P]

A sequence of numbers in which each number is the sum of the previous two, beginning 0, 1, 1, 2, 3, 5, 8, 13, 21… . A sunflower's seeds are arranged in Fibonacci number patterns.

Harmonics [P]

Musical instruments rarely play pure notes, producing harmonics – different pitched notes that come together to produce the instrument's distinctive sound. Harmonics are the reason why the same note sounds very different on the saxophone and on the keyboard.

Longitudinal waves [P]
Waves that vary cyclically in the direction of travel, rather than at right angles to it. The sound is carried through the air by longitudinal waves as the air squashes and expands like a concertina.

Moon illusion [P]
An optical illusion makes the Moon appear bigger than it really is, particularly when it is near trees, buildings or the horizon. This is why the Moon looks surprisingly small in photos of it in the night sky.

Lotus effect [P]
Some natural materials have self-cleaning properties because the surface of the material repels water, forcing the water into droplets that carry dirt away with them. Water lily leaves exhibit the lotus effect.

Negative pressure [P]
When one gas pressure is lower than another, the relative pressure can be expressed as a negative value. When the boy sucks, the pressure at the top of the straw is less than the air pressure on the surface of the liquid, so this negative pressure results in the drink rising up the straw.

Low-light vision [P]
In low light, the more sensitive cells in the eye called rods dominate. Unlike the colour-sensing cones, rods only detect monochrome, so it's not possible to tell what colour the apple is in the near-dark.

Noctilucent clouds [P]
Clouds that are high in the sky can still be illuminated by the Sun when it is hidden from sight by other clouds or the horizon. Clouds like these are called noctilucent.

Mendelian inheritance [P]
The concept of genetic inheritance based on the characteristics of parents. Pea plants like these were used by Gregor Mendel in experiments to determine the effects of Mendelian inheritance.

Nyctinasty [P]
Movement in plants as darkness falls. Some flowers' petals, such as these, close up at night, exhibiting nyctinasty. It is not clear how this benefits the plant.

Osmosis [P]
The movement of liquid from one side of a membrane to another, travelling to the side that has a higher concentration of dissolved substances. When the plants are watered, they absorb water from the soil by osmosis.

Phototaxis [P]
When an organism moves towards or away from light. The moths are attracted by the light of the candle flame, as they naturally use the light of the Moon to navigate.

Ouzo effect [P]
The production of a cloudy liquid when water is added to some oils dissolved in alcohol, because of the formation of an emulsion (a mix of fine droplets of oil in the water). As a result of the ouzo effect, the pastis becomes white as water is added.

Plasma [P]
The fourth state of matter after solids, liquids and gasses is plasma. This is similar to a gas but made up of charged particles called ions, which are atoms that have lost or gained electrons. The flame of the candle is rich in plasma.

Parallax [P]
The difference in apparent movement of objects at different distances. As the children run along, it feels as if the Moon is following them, because the nearer trees and fence seem to move past more quickly.

Polarising filter [P]
A polarising filter cuts out light that has been polarised (see Brewster's law, page 32) in a particular direction. The polarising filters in the woman's sunglasses are aligned to reduce levels of reflected light.

Photoelectric effect [P]
The generation of an electrical current from a semiconductor or metal when light falls on it. The night vision goggles take in both visible and infra-red light and use this effect to generate electrical signals, which are converted to visible light, making it possible to see in apparent darkness.

Self-organised criticality [P]
Some systems naturally reach critical points at which sudden change occurs. The pile of sand has self-organised criticality – as grains of sand are added there will be a sudden collapse when the pile becomes too high.

Self-similarity [P]
Structures known as fractals have self-similarity: when examined closely, the detail resembles the whole structure. A fern has self-similarity as fern fronds resemble fern plants.

Supermoon [P]
When the Moon is full at the point when its orbit is closest to the Earth, it appears around 14% bigger and is called a supermoon. The supermoon magnifies the Moon illusion (see page 35).

Spectroscopy [P]
When a substance is heated, the chemical elements present can be identified by the colours of light given off. It is such spectroscopic effects that enable firework makers to produce different coloured flames by using appropriate chemical compounds.

Surface tension [P]
Water molecules are attracted towards each other, which means that without obstacles they tend to form spherical drops. The attraction the molecules on the surface feel towards the rest of the water is called surface tension. The water drop forms on the girl's nose through surface tension.

Standing waves [P]
Some structures allow the build-up of waves of particular frequencies which do not move along but are held in place by restrictions in the environment. The musician's saxophone produces musical notes depending on standing waves that can form in its tube.

Ultra-hydrophobicity [P]
Some objects are particularly good at repelling water, known as ultra-hydrophobicity. Large numbers of hairs on the legs of the water strider beetle repel water and help it to walk on the surface.

Sunset [P]
As the Sun gets lower in the sky, its colour shifts from yellow-white to red. This is because the light passes through more air, which tends to scatter blue light more. Sunset's red-orange glow is the result.

Wavelength of waves [P]
A wave is defined by its amplitude (size) and its wavelength (distance between similar points on the wave). Because the ends of the dog's rope are fixed, it limits the possible wavelengths that can be produced by plucking it, and hence determines the note it makes.

The Science Museum

The Science Museum

Born's law [L]
In quantum physics, determines the probability of finding a particle in a location. Mirrors with missing sections change the probability of light particles reflecting, so light bounces off at an unexpected angle.

Fermi's golden rule [L]
Describes the probability an electron in a semiconductor will lose energy, giving off a photon of light. This determines the brightness of LED bulbs.

Chemical periodicity [L]
The reactions of an element depend on the number of electrons at the atom's edge: electrons build up in layers called shells, producing a regular pattern of similar elements. This shapes the periodic table.

Pauli exclusion principle [L]
No two electrons in a system can be in the same state (for example, with the same position or energy). This principle shapes the behaviour of the computer chip, shown in the circuit diagram on the wall.

Conservation of charge [L]
The total charge in a system stays the same. Van der Graaf generators build up charge on the dome by transferring electrons from the metal on to a rubber belt.

Standard model of particle physics [L]
The standard model describes 17 particles responsible for matter and all physical forces except gravity. In the particle accelerator collision, many particles are created.

Third law of thermodynamics [L]
The third law says it's impossible to reach absolute zero (−273.15°C / −459.67°F). The cooling device can get extremely close to absolute zero, but never reaches it.

Atomic structure [P]
Atoms have small dense central nuclei and electrons in a fuzzy cloud outside, being mostly empty space. The 'solar system' diagram is inaccurate, because electrons do not orbit like planets, but it's a familiar representation.

Uncertainty principle [L]
Law of quantum physics linking pairs of properties, such as energy and time. The more accurately we know one, the less we know the other. Metal plates under the microscope are pushed together by particles briefly created by energy fluctuations.

Bose–Einstein condensate [P]
This special state of matter slows photons to a crawl, or traps them. In the experiment, light is temporarily held in a Bose–Einstein condensate.

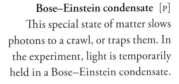

Amorphous solid [P]
Many solids are crystals, but some have a messy structure without patterns. Glass is a classic amorphous solid.

Carbon dating [P]
Carbon has a radioactive isotope, carbon-14. This decays over time, so the amount present shows when it formed. The accelerator mass spectrometer measures the carbon-14 present.

Atomic nucleus [P]
Most of an atom's mass is concentrated in the nucleus. In this experiment, particles are fired at gold foil. Some of the particles bounce back, showing the existence of the nucleus.

Casimir effect [P]
The uncertainty principle means that particles pop briefly in and out of existence in empty space. The resultant pressure forces two very close flat objects together in the 'Casimir effect' – the metal plates underneath the microscope demonstrate this effect.

Cladistics [P]
This classifies biological species based on their genetic common ancestors. The diagram shows where different species split off from common ancestors.

Doppler cooling [P]
Extremely cold temperatures, such as those needed for zero resistance (see page 47), are produced using lasers in Doppler cooling. Light is absorbed by atoms, slowing them and reducing temperature, which is a measure of atomic speed.

Crystalline solid [P]
Many solids are crystals with atoms linked together in regular, repeating patterns. Carbon has a number of crystalline forms including shiny black graphite.

$E=mc^2$ [P]
Einstein showed that matter and energy are interchangeable with the relationship described by this equation, where E is energy, m is mass and c is the speed of light.

Dendrochronology [P]
Each tree ring represents a year's growth, so counting rings dates the wood, which is used to calibrate carbon dating. The rings get older from outside to centre.

Graphene [P]
Graphene is a single atom-thick layer of graphite with special quantum properties, making it extremely conductive and strong. Graphene was first produced using sticky tape to remove thin layers from graphite.

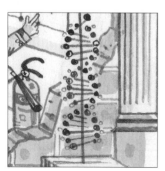

DNA structure [P]
A crucial part of understanding the role of DNA was working out its structure. The model shows the spiral staircase-like dual helix structure of DNA.

Holograms [P]
A three-dimensional image, produced by scanning with a pair of lasers. The tiger is flat, but looks three-dimensional with the appropriate illumination.

Many worlds hypothesis [P]
Attempt to explain the oddities of the quantum, suggesting that each time there is more than one possible outcome, each occurs in a different universe: in one universe the cat is dead, in another alive (see Schrödinger's cat, page 46).

QED [P]
Quantum electrodynamics (QED) is the science of the interaction of light and matter. The Feynman diagram shows the interaction of particles of matter and light.

Meissner effect [P]
Some materials become superconductors with no electrical resistance when cooled near absolute zero. They repel magnetic fields, so a magnet floats over them in the Meissner effect.

Quantum entanglement [P]
A phenomenon where quantum particles interact instantly at any distance. Quantum entanglement can be used to distribute random values that encrypt secret messages.

Metamaterials [P]
Special materials with a negative refractive index bend light the opposite way to water. Such metamaterials are used in specialist lenses and invisibility cloaks, which bend light around objects.

Quantum spin [P]
Quantum particles have a property called spin. This does not involve spinning: when measured it can only be in two directions, up or down. The children are looking at the Stern–Gerlach experiment, detecting quantum spin.

Planck's constant [P]
A constant of nature linking a photon's energy to its frequency (colour). The photoelectric effect, used in digital cameras, solar panels and the light detector above the door, detects colours from the photons' energy.

Quantum superposition [P]
A quantum particle has probabilities of being in different states before being measured. This is a superposition of states. The Schrödinger's cat experiment (see page 46) is controlled by a particle in a superposition of being decayed and not decayed.

Quantum tunnelling [P]
A quantum particle can pass through a barrier that should stop it because there is a probability it is already on the other side. In the experiment, light tunnels through the gap between two prisms.

Schrödinger's cat [P]
Here (theoretically), a cat is placed in a box with a radioactive particle, a detector and a vial of poison. When the particle decays it triggers the poison, killing the cat. As the particle is both decayed and not, the cat is both dead and alive.

Quark confinement [P]
Quarks, the particles in protons and neutrons, are attracted by the strong force, which gets stronger as quarks separate. Quarks are 'confined': not seen on their own. Particle accelerators use extreme energy to overcome confinement.

Schrödinger's equation [P]
This equation shows the probability of finding a quantum particle at different locations. Because the particle's location is represented by a probability wave, it can effectively pass through two slits and interfere, causing a pattern of light and dark areas.

Radioactive decay [P]
The nuclei of some atoms are unstable, breaking apart to produce multiple particles in radioactive decay. This is the source of nuclear radiation, and triggers the outcome of the Schrödinger's cat experiment.

Special theory of relativity [P]
Einstein's theory connecting time and space shows that as objects get faster, time slows and mass increases. In the accelerator, particles near light speed and these effects become noticeable.

Refraction [P]
The path of light bends when it passes between materials where it has different speeds. The pencil appears to bend where it passes from air to water because light is slower in water.

Speed of light [P]
The speed of light is constant in a medium. This is measured using a laser and a detector: light passes from laser to detector at around 299,700 kilometres (186,000 miles) per second.

Stimulated emission of radiation [P]

A laser boosts the energy of an electron in an atom using a photon of light, then triggers the release of that energy with a second photon, amplifying the light (Light Amplification through the Stimulated Emission of Radiation).

Van der Graaf generator [P]

A device producing high-voltage electricity. Touching the dome of the generator transfers the electrical charge from person to person, making hair stand on end.

Superfluid [P]

Some liquids become superfluids with no viscosity close to absolute zero. Once moving, they don't stop, and flow out of containers. The narrow opening produces a self-powered fountain.

Viscosity [P]

Viscosity measures gooiness. One of the most viscous substances is pitch: the pitch drop experiment, which has been run for over 90 years, has only dripped eight times.

Superluminal speeds [P]

Because light tunnelling through a barrier does so near-instantaneously, it travels faster than light. The light passing through the prisms reaches about four times light speed.

Wave/particle duality [P]

Quantum particles exhibit wave-like behaviour. The dual-slit experiment works with electrons, fired one at a time. As with light, the result is interference and a pattern of fringes builds up.

Tesla coil/induction [P]

High-voltage alternating current induces a strong electrical current nearby. The woman holds a fluorescent tube near the high-voltage source and the tube glows.

Zero resistance [P]

Some materials cooled near absolute zero lose all electrical resistance, becoming superconductors. An electrical current continues indefinitely, and is used to make ultrapowerful magnets. The meter displaying the electrical current goes off the scale.

The Hospital

The Hospital

Poiseuille's law [L]
Describes the pressure change as a fluid flows through a cylinder significantly longer than its width, such as a hypodermic needle.

Angiogenesis [P]
The process by which new blood vessels form. As the injury on the man's leg heals, angiogenesis is an essential part of the wound healing.

Aerobic/anaerobic exercise [P]
Aerobic exercise, such as running, generates energy from carbohydrates using oxygen; anaerobic exercise generates energy from glucose without oxygen. The equipment monitors aerobic exercise.

Antimatter [P]
In the PET (positron emission tomography) scanner, radioactive material injected into the patient gives off antimatter (positrons), which interacts with electrons to generate gamma rays, detected by the PET scanner.

Anaesthesia [P]
A reduction of a patient's awareness or consciousness to enable medical procedures to be carried out without pain. Can be administered by gas, injection or orally.

Antiperistalsis [P]
Also called retroperistalsis, a reversal of peristalsis. Waves of muscle action move food through the system, returning material to the stomach prior to vomiting.

Auscultation [P]
Using sounds from inside the body to diagnose medical conditions. The stethoscope, bringing sounds to the doctor's ears, is the usual tool for auscultation.

Digestion [P]
Breaking down large food molecules into smaller ones that are usable by the body. Chemical processes in the patient's digestive system break down his food.

Blood typing [P]
Blood is grouped into types depending on antibodies present. The test undertaken here is essential to ensure appropriate blood types are used for safe transfusion.

DNA fingerprinting [P]
Also known as DNA profiling, a mechanism for comparing DNA samples to identify forensic material and determine paternity. Produces patterns that can be compared.

CRISPR [P]
A technique enabling precise edits of the DNA of living organisms, including humans. It is likely to be widely used in treating human genetic diseases.

DNA replication [P]
The dual helix of the DNA molecule containing genetic information splits into two halves, each containing the full information, allowing DNA to be copied when cells divide.

Dialysis [P]
A mechanism for removing excess water and toxins from the blood. The dialysis machine replaces the action of the patient's failing kidneys.

Electrocardiography [P]
An electrocardiograph (ECG) machine uses electrical contacts on the skin in order to measure the electrical activity of the heart, detecting problems with heart rhythm and function.

Electroencephalography [P]
An electroencephalogram (EEG) machine uses readings from electrodes on the scalp to detect electrical activity in the brain. This can be used to diagnose epilepsy and other brain conditions.

Endosymbiosis [P]
Organisms that work within others to mutual benefit. Mitochondria, small modules in cells that process ATP energy-store molecules, were once bacteria which evolved an endosymbiotic relationship.

Epigenetics [P]
Genes only form a small percentage of DNA. Much of the rest incorporates mechanisms to switch genes on and off. The science of this non-gene DNA is epigenetics.

Eukaryotes [P]
The type of cell found in animals, plants and fungi, including us. Eukaryotic cells have a central nucleus containing much of the molecular machinery.

Flagellum [P]
Many bacteria have a built-in molecular motor, used to move an external whip-like structure known as a flagellum, propelling the bacterium along.

Genes [P]
Sections of the DNA molecule containing the information required to produce particular molecules needed for an organism to function, notably proteins.

Haemostasis [P]
The body's mechanism to stop bleeding, whereby the blood coagulates to form a gel. This is the first stage of a wound healing.

Homeostasis [P]
The process regulating a system, as a thermostat does. Mammals, including humans, use a number of homeostatic mechanisms to keep the body at constant temperature.

Hypertension [P]
The condition where arterial blood pressure is above normal. The cuff of the blood pressure monitor inflates then slowly deflates to measure maximum and minimum pressure.

Meiosis [P]
The genetic process where an individual's chromosomes (molecules of DNA-holding genes) are split and remixed to form the genetic material in sperm or eggs.

Inflammation [P]
The body's response to infection or injury – causing pain, redness and swelling as the immune system attempts to eliminate the problem and start repairs.

Metabolism [P]
A collective term for the processes that provide the energy and remove waste from a biological organism. The food that the patient is eating will fuel his metabolism.

Infusion [P]
Infusion pumps add fluids, including medication, to the patient's bloodstream or subcutaneously. They are more effective than injections for small or regularly administered doses.

Mitosis [P]
The process by which a eukaryotic cell's nucleus divides into two, separating already duplicated chromosomes, so that the cell can divide to produce two cells.

Krebs cycle [P]
The mechanism in mitochondria, also called the citric acid cycle, converting energy sources such as carbohydrates into intermediaries used to generate the energy-store molecule ATP.

**Morphogenesis –
Turing patterns** [P]
As embryos grow, their development of form is called morphogenesis. Alan Turing demonstrated how interacting fluids can produce a regular pattern, responsible for some morphogenetic structures.

Neurotransmission [P]
The brain operates when electrochemical links pass messenger chemicals between neurons, known as neurotransmission. The MRI (magnetic resonance imaging) scanner can identify neural problems.

Prokaryotes [P]
Single-celled organisms, such as bacteria and archaea, which do not have a nucleus. Understanding the cell structure of prokaryotes helps in antibacterial work.

Nucleus [P]
A structure inside a eukaryotic cell surrounded by a membrane. The nucleus houses the cell's chromosomes, each a long, single molecule of DNA.

Protein synthesis [P]
The process by which proteins – large organic molecules with a wide range of uses – are constructed. Understanding protein structures is central to molecular biology.

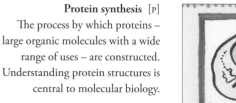

Phage [P]
Short for 'bacteriophage', a phage is a virus that attacks bacteria, often looking strangely like a Moon lander. Phages are being considered as alternatives to antibiotics.

Proton pump [P]
An important biological mechanism that moves electrically charged protons through membranes. The concentration of electrical charge is used as a way to store energy.

Photoreception [P]
The mechanism by which special cells in the eyes detect light. The medical professional uses an ophthalmoscope to examine the retina, which houses photoreceptors.

Reflexes [P]
A localised stimulus and reaction that does not require intervention by the brain. By tapping the knee, a doctor can check that the patient's nerves are functioning correctly.

Respiration [P]
Respiration brings oxygen into the body to react with nutrients and carries away carbon dioxide. The oxygen cylinder provides extra oxygen when the body is struggling.

Vaccination [P]
Introducing materials into the body that encourage the immune system to build a natural protection against infection. Important to develop protection against viruses, which are not susceptible to antibiotics.

Superconductivity [P]
The MRI scanner uses extremely powerful superconducting magnets, which work by using a quantum effect where electrical resistance drops to zero at ultra-low temperatures.

Virus [P]
A tiny cell-like object that causes infections. Unlike bacteria, which are standalone living organisms, viruses use the mechanism of their hosts' cells to replicate.

Thixotropy [P]
A thixotropic fluid flows more easily when it is shaken or stressed. Ketchup is thixotropic, as is the non-drip paint used on the wall.

Weak interaction [P]
One of the four fundamental forces of nature, controlling nuclear decay. The PET scanner picks up radiation from a radioactive material introduced into the patient.

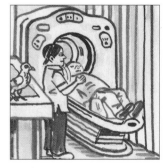

Ultrasound [P]
Sound that is too high pitched to be audible. The scanner emits ultrasound, which bounces off internal structures, such as a foetus, to produce an image.

X-rays [P]
High-energy form of light, outside the visible range, which passes through flesh but is stopped by bones, enabling internal examination of a patient.

The Town Square

The Town Square

Ampère's law [L]
Relates the magnetic field around a closed loop to the electric current passing through it. The changing magnetic field in an attached loop of wire moves the cone of the portable stereo's loudspeaker.

Bernoulli principle [L]
An increase in speed occurs with a decrease in pressure or potential energy. The paper planes are lifted by changes in pressure on the surfaces as air flows around them.

Archimedes' law of the lever [L]
Forces are in equilibrium at distances reciprocally proportional to their magnitudes. Archimedes' law explains how the long lever enables the man to jack up his car.

Boyle's law [L]
The pressure and volume of a gas are inversely proportional to each other, when temperature is constant. This allows the bicycle pump to work.

Archimedes' principle of flotation [L]
A body completely or partially submerged in a fluid feels an upward force equal to the weight of the fluid displaced by the body: the helium balloons displace the heavier air and therefore float.

Brewster's law [L]
The polarisation of a reflected ray of light is dependent on the angle at which it enters a transparent medium. The reflected sun's rays are polarised by the mirror.

Charles' law [L]
The volume of an ideal gas is proportional to its temperature at constant pressure – as the air is heated it expands, making the balloon less dense. It flies by Archimedes' principle (see opposite).

Henry's law [L]
The amount of gas dissolved in a liquid is proportional to the partial pressure of the gas above the liquid. Watch out for the champagne cork!

Conservation of angular momentum [L]
A change in angular momentum is proportional to the applied torque (twisting force) and occurs about the same axis as that torque. With the handlebar straight the bike still turns.

Hooke's law [L]
The force needed to extend or compress a spring (elastic material) is proportional to the expansion or compression. That's how the little boy can play with his jack-in-a-box.

Faraday's law of induction [L]
Predicts how a changing magnetic field induces a current in a nearby electric circuit. This powers the transformers, charging the boys' smartphones.

Joule heating [L]
The amount of heat released when electric current flows through a material is proportional to the square of the current. The hairdryer produces hot air from a wire heated by electricity.

First law of thermodynamics [L]
States that the total energy of an isolated system remains constant (energy cannot be created or destroyed). Chemical energy in the oil is converted to heat by burning.

Law of radioactive decay [L]
Predicts how the nuclei of a radioactive substance decay over time. Bananas are radioactive, with potassium atoms spontaneously decaying to become calcium. Don't worry, they are not dangerous.

Leighton relationship [L]
Predicts the concentration of ozone in the lowest layer of the atmosphere by the nitrogen oxides present, since ozone is produced by the photolysis ('light-breaking') of nitrogen oxides.

Newton's first law [L]
The law states that an object stays at rest or in constant motion unless a force acts upon it. With no applied force, the parked car does not change its (lack of) motion.

Newton's law of cooling [L]
The rate of heat loss is proportional to the difference in temperatures between a body and its surroundings. The soup cools quickly in the late afternoon chill.

Newton's law of gravitation [L]
Two bodies attract each other with a force proportional to the product of their masses and inversely proportional to the square of the distance between them. Watch out down below!

Newton's second law [L]
The net force on an object is proportional to its acceleration. The skateboarder pushes off with one foot to speed herself up.

Newton's third law [L]
For every action there is an equal and opposite reaction. The dog pulls on the lead while his owner stands still. The tension in the lead pulls back on the dog.

Planck's law [L]
Describes how the colour of light emitted varies with temperature. The red-hot end of the woman's cigarette radiates infra-red energy according to Planck's law.

Second law of thermodynamics [L]
Heat cannot spontaneously flow from a colder to a hotter location. The ice lolly does not make the air surrounding it warmer.

Capillary action [P]
The ability of a liquid to flow in narrow spaces without or against forces such as gravity. This enables the waiter to soak up a spilled drink with the towel.

Doppler effect [P]
The change in frequency of a wave when the source is relative motion. The pitch of the ambulance's siren changes as it speeds past people.

Cooling by evaporation [P]
The energy required to detach molecules from a liquid to evaporate it results in a drop in temperature. The fan encourages evaporation, cooling the skin.

Dynamic friction [P]
Friction between solid objects that are moving relative to each other. The boy is braking too hard, causing excess friction from the brake block.

Crepuscular rays [P]
Rays of sunlight from the point in the sky where the Sun is located. The rays of sunlight are parallel, but an optical illusion makes them appear to radiate.

Elastic materials [P]
When elastic is stretched it unkinks twisted molecules, which constantly pull back against the stretching. The child's rein is an elastic material.

Diffusion [P]
The movement of molecules from a higher to a lower region of concentration until they are equal. The dogs smell molecules from the food, diffusing through the air.

Electroluminescence [P]
When a material emits light in response to the passage of an electric current or to an electric field, as in the LEDs used in traffic lights.

Entropy [P]
A measure of the degree of disorder in a system. Entropy tends to stay the same or increase. It is easier to break the bottle than to unbreak it.

Joule–Thomson effect [P]
The temperature change of a fluid forced through a narrow opening while insulated so no heat is exchanged with the environment, used in the ice-cream cart.

Hydrogen bonding [P]
The electrical attraction between the hydrogen atom in a molecule and an atom such as oxygen in a separate molecule. Hydrogen bonding keeps the water in the chef's pot liquid.

Marangoni effect [P]
Mass transfer along an interface between two fluids. The 'tears of wine' in the brandy glass occur because of alcohol's lower surface tension than water.

Interference [P]
The process by which two waves interact, reinforcing or cancelling each other out. The waves in the pond interfere.

Mechanical advantage [P]
A measure of the amplification of force produced by a machine. The bike's gears amplify the force put on the pedals.

Iridescence [P]
The production of colours from surfaces as the angle of view or illumination changes. Bands of colour form in the thin film of oil on the puddle.

Melanin and ultraviolet light [P]
Melanin is a natural pigment, produced in the skin on exposure to ultraviolet light. The woman's tan shows an increase in melanin.

Oxidation [P]
The gain of oxygen, or more generally the loss of electrons from an element or compound. The combustion of the fire is a dramatic form of oxidation reaction.

Rayleigh scattering [P]
The scattering of light or other electromagnetic radiation by particles smaller than the wavelength of the radiation. The small gas molecules of air particularly scatter blue light, giving the sky its colour.

Particulates [P]
Small particles, which can be suspended in the air as an aerosol. The particulates from the ambulance's exhaust form an aerosol.

Resonance [P]
The reinforced vibrations in a system when agitated near its natural frequency. The tuning forks resonate to the frequency of the trumpet note coming from the room above.

Photosynthesis [P]
The process by which plants and some bacteria use the energy from sunlight to produce glucose from carbon dioxide and water.

Secondary cosmic ray shower [P]
Particles that are produced when high-energy cosmic rays collide with molecules high in the atmosphere. Most of these particles are muons, which subsequently decay, producing many more particles.

Photovoltaic effect [P]
The creation of voltage or electric current in a material upon exposure to light, as in this solar panel.

Static friction [P]
Friction between two or more solid objects without relative motion. The static friction between the woman's knees and the plank prevents her from slipping.

The Street

The Street

Boyle's law [L]
Expanding a body of gas reduces pressure. In the car's engine, the pressure drops as the piston moves in the cylinder.

First law of thermodynamics [L]
Energy is conserved but can change form. When the woman lifts the boxes, the work done converts to potential energy and heat.

Charles' law [L]
A gas expands as it is heated. In the bike tyre, the air expands as friction with the road heats up the tyre.

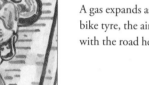

Gay-Lussac's law [L]
For a constant volume of gas, increasing temperature increases pressure. In the starting pistol, gunpowder rapidly increases temperature and hence pressure, producing a bang.

Faraday's law of induction [L]
The law showing how electricity is induced by magnetism. Some electric cars use an alternating current induction motor based on Faraday's law.

Lambert's first law [L]
The illuminance on a surface is proportional to the inverse square of the distance from the source. It's hard to read with limited lighting.

Lambert's second law [L]
Illuminance is dependent on the angle at which light hits an object. The brightness of the map depends on the angle at which it's held.

Amphiphilic substances [P]
Such substances are attracted to both water and fats. The detergent used to clean the window is amphiphilic.

Lambert's third law [L]
The intensity of light decreases exponentially with distance as it travels through an absorbing medium. The inside of the shop looks dim behind thick glass.

Anchoring effect [P]
A cognitive bias where an individual is over-dependent on an early piece of information. Although $9.99 is almost $10, the shopper is influenced more by the $9 part.

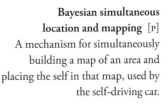

Snell's law of refraction [L]
The change of direction when light moves from one medium to another is fixed for those media. The woman's sunglasses bend light moving from air to glass.

Bayesian simultaneous location and mapping [P]
A mechanism for simultaneously building a map of an area and placing the self in that map, used by the self-driving car.

Zipf's law [L]
The frequency of word use in a language (and this conversation) is inversely proportional to a power of their ranking – the frequency of a word decreases significantly as the rank number increases.

Beam splitter [P]
Beam splitters allow some light through, but also send light in a different direction. The one-way window glass is like a mirror as the interior is darker than the street.

Cosmic rays [P]
A stream of particles hitting the Earth from deep space. Hundreds of cosmic ray particles pass through the woman's body every second.

General theory of relativity [P]
Einstein's theory linking gravity to warps in space-time shows that gravity slows time. The GPS system has to correct for lower gravity where its satellites orbit.

Diffraction [P]
Waves bend on hitting an obstacle, making it possible for the boys to hear the discussion around the corner of the building.

Gyroscopic effect [P]
A spinning disc resists movement away from its direction of spin. The rotating wheels of the bicycle contribute to the bike's stability when riding hands free.

Electromagnetic absorption [P]
When light passes through a coloured transparent material, photons of some energies are absorbed. The colour of the traffic lights depends on photons that are not absorbed.

Hall effect [P]
A voltage produced at right angles to the flow of current through a conductor in a magnetic field. This is used in an electronic ignition timing device in a car.

Gas discharge [P]
A type of lighting producing light by sending an electrical current through a charged gas. The neon signs are gas-discharge lamps.

Infra-red laser [P]
A laser producing light in the infra-red spectrum, lower in energy than visible light. The fibre optic communication cable carries infra-red laser light.

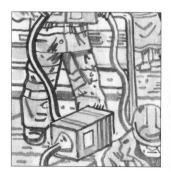

Lidar [P]
Like radar, but using lasers to measure distances to surrounding objects. Used for collision avoidance in most self-driving cars.

Quantum biology [P]
Biological processes using quantum effects. Pigeons' ability to navigate using the Earth's magnetic field appears to be a quantum phenomenon.

Machine learning [P]
Computer programs that can modify how they behave dependent on data that they process. Used to control self-driving cars.

Radio frequency identification [P]
Radio waves from the security system in the doorway of the shop induce current in tags on the products, sounding the alarm.

Mechanical advantage [P]
The degree to which a force is amplified by a machine. Pulleys on the window-washing cradle provide mechanical advantage.

Regelation [P]
The process of melting under pressure. This requires a lot of pressure: the slipperiness of ice is caused by loose water molecules on the surface, not regelation.

Power [P]
The power of vehicles is measured in horsepower, the equivalent of 736 watts. This is roughly the steady power output of a horse.

Resonance [P]
Large oscillations when an object is vibrated at close to its natural frequency. Buses sometimes shudder wildly when their engines run at their resonant frequency.

Retroreflective materials [P]
Special materials that reflect most incoming light back without scattering it. Modern bike reflectors are retroreflective.

Speed of light in material [P]
Light travels slower in air than in a vacuum, and slower in glass than air. The light in the fibre optic cable travels around 200,000 kilometres (124,000 miles) per second.

Self-organising system [P]
A system that naturally organises itself in a particular way. Snowflakes self-organise to be six-sided because of the shape of the water molecule.

Thin film interference [P]
Light reflecting off a thin film of liquid can interfere with light reflecting through it. This produces rainbow colours in the thin film of oil underneath the car.

Sonic boom [P]
The noise from a supersonic plane. The aircraft produces a loud bang when the distance to the ear is just right for sound waves to reinforce each other.

Torque [P]
A force applying rotational motion to an object. The motorbike rider leans into the turn in order to balance the tyre friction and the force producing the turn.

Special theory of relativity [P]
Einstein's theory linking space and time, resulting in time slowing on moving objects. The GPS satnav corrects for this in the motion of its satellites.

Total internal reflection [P]
Light hitting the boundary between a substance and something less dense at a shallow angle stays inside. This keeps laser light in the fibre optic.

Traction [P]

The adhesive force of friction between two surfaces. The tread on the car tyres increases the area in contact and the level of traction.

Van der Waals force [P]

Electrostatic adhesive force between atoms or molecules. This is how geckos are able to walk up walls and is used in special gloves and pads in order to climb glass.

Triangulation [P]

Fixing a location in three dimensions by measuring the distance to three known points. The GPS app on the smartphone uses triangulation.

Vortices [P]

The rotating whirls of a fluid. The flag is flapping in the wind as vortices form and collapse.

Tyndall effect [P]

A suspension of particles in a transparent medium scatters blue light more than other frequencies. This is why a motorbike's exhaust can look blue.

Vulcanisation [P]

Process using sulphur and other substances to harden rubber. Car tyres are manufactured using vulcanised rubber.

Urban heat island effect [P]

In towns, paved areas and buildings store heat, which prevents the temperature from falling at night as much as it otherwise would.

Wind tunnel effect [P]

Moving air speeds up as it passes from an open space into a narrower gap to get the molecules through. The wind tunnel effect blows the man's hat off.

The Countryside

The Countryside

Cassie's law [L]
Describes the angle of the edge of liquid in contact with a substance of a different chemical composition. Water droplets on the duck's feathers follow this law.

Commoner's first law [L]
Ecological law stating, 'everything is connected to everything else'. The smoke from the factory has an impact on the wider environment.

Commoner's second law [L]
According to this ecological law 'everything must go somewhere'. The landfill waste isn't thrown away – it remains part of the environment.

Kleiber's law [L]
In most animals, energy consumption goes up with approximately the ¾ power of weight. About 50 times heavier than the rabbit, the wolf uses 19 times the energy.

Stokes' law [L]
Describes the force on spherical objects moving smoothly in a fluid. Clouds, made up of tiny droplets, fall under gravity very slowly due the strong drag predicted by Stokes' law.

Aerobic respiration [P]
Energy production in cells that makes use of oxygen. The runner's steady exercise produces aerobic respiration.

Asexual reproduction [P]
Reproducing without multiple sexes. Ferns have a complex reproductive pattern that is partly asexual.

Clonal colony [P]
A group of organisms that grow together as clones. Hazelnut trees often grow close together in this way, connected to the same root system.

Bioluminescence [P]
The ability some organisms have to produce light by chemical processes. The firefly makes use of bioluminescence for signalling.

Colour vision [P]
Some animals have different colour vision ranges to humans. Using ultraviolet colour vision the kestrel has the ability to spot a mouse by seeing its urine trail.

Bruce effect [P]
Some female rodents terminate pregnancies when exposed to the scent of an unfamiliar male. The effect is best known in mice.

Contrails [P]
Also known as vapour trails: these are linear clouds formed when water from a jet engine's exhaust is cooled by low air temperatures.

Butterfly effect [P]
An implication of chaos theory that a small change in circumstances will have significant consequences. The original example given was the flap of butterfly wings causing a tornado.

Convergent evolution [P]
The independent evolution of features with the same function in different organisms. Insect eyes evolved separately from those of birds.

Ectothermy [P]
Organisms where body temperature is primarily controlled by outside temperatures. Lizards and other such ectothermic animals are often misleadingly called cold-blooded.

Evapotranspiration [P]
The evaporation of water from the land and from plants after it emerges from their leaves. Evapotranspiration from the leaves increases humidity.

Edge effect [P]
Ecological term describing the situation at a boundary between habitats. Due to the edge effect there is increased biodiversity in the region between woodland and grassland.

Fractal nature [P]
Fractals are mathematical structures that are self-similar, where parts of the structure resemble the whole. Some trees, particularly conifers, are examples of fractals in nature.

Electromagnetic repulsion [P]
The repulsion between particles with the same charge due to the electromagnetic force. This electromagnetic repulsion between the atoms in bricks keeps the building stable.

Gravity [P]
Gravity produces potential energy when an object is higher than another location. The river, powering the water wheel, runs downhill due to gravity.

Endothermy [P]
Organisms where body temperature is controlled by internal processes. Mammals and birds, for example, are endotherms, sometimes described as warm-blooded.

Hibernation [P]
The tendency of some endothermic animals to enter a state of low metabolic activity in order to survive the winter. Hedgehogs are familiar hibernators.

Hydrogen bonding [P]
Attraction between relatively positive and negative parts of molecules, such as between hydrogen and oxygen, increasing boiling point. This hydrogen bonding keeps the lake's water in a liquid state.

Metamorphosis [P]
A rapid change in form of an animal, often involving significant transformation of cells. The caterpillar undergoes metamorphosis to become a butterfly.

Ideomotor effect [P]
The ability to produce muscle movements without conscious effort. Dowsing rods move because of the ideomotor effect.

Murmuration [P]
The collective movement of a flock or shoal of animals, where each animal is influenced by others nearby. Starlings produce dramatic murmurations.

Imprinting [P]
Rapid learning at a specific stage of life. Young birds often imprint to follow their parent: geese follow a microlite after they have imprinted on it as a parent.

Natural selection [P]
Major mechanism of evolution where organisms better able to survive in an environment reproduce. Moths that more closely resemble tree bark are more likely to avoid being eaten.

Logistic equation [P]
Describes the growth of a population, based on the number the environment can sustain. The population of rabbits here is controlled by the logistic equation.

Navier–Stokes flow [P]
The Navier–Stokes equations describe the steady flow of a liquid without turbulence. The smooth-flowing part of the stream has Navier–Stokes flow.

Nitrogen fixation [P]
Plants use nitrogen from the air in their makeup. The roots of some plants is where fixation occurs, assisted by bacteria that live among the roots.

Pollination [P]
Mechanism by which pollen is moved from male to female plant parts. A number of insects, notably bees, carry pollen from plant to plant as they collect nectar.

Nocturnal eyesight [P]
The ability to see in low light. The owl's eyes are tubular, meaning it has to rotate its head to extreme angles to see in different directions.

Rabbit breeding [P]
The reproductive habits of rabbits were used in the thirteenth century by Fibonacci to illustrate the sequence of numbers now known as the Fibonacci series.

Photonic lattice [P]
A structure producing optical effects at the quantum level. The iridescence of the butterfly's wings is produced by a photonic lattice.

Rainbow [P]
An optical effect producing a spectrum of colours when white light from the Sun is split out into its constituent colours by passing through raindrops.

Photosynthesis [P]
The conversion of light to chemical energy. Plants produce energy from sunlight by photosynthesis.

Respiration [P]
Controlled combustion, used to produce energy from chemical bonds in food by living organisms. The squirrel's nut produces energy through cellular respiration.

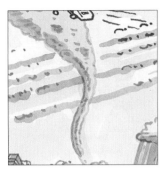

Self-organising system [P]
The capability of a system to spontaneously produce its own structure by local interaction of its parts. The tornado is an example of a self-organising system.

Terminal velocity [P]
The speed at which the resistance of a fluid stops a falling body from accelerating. The parachute reduces terminal velocity, keeping the parachutist safe.

Sexual reproduction [P]
The reproduction of organisms as a result of combination of genetic material from a male and female. Like all mammals, the rabbits reproduce sexually.

Trophic cascade [P]
When a predator reduces the population of a lesser predator, reducing predation of the prey of the lesser predator. The wolf can cause a trophic cascade.

Strong interaction [P]
The mechanism that holds atomic nuclei together. Some of this energy is released in radioactivity, such as the background radioactivity from these granite rocks.

Turbulent flow [P]
When a fluid's flow becomes chaotic, producing unpredictable sudden changes in pressure and flow rate. The stream undergoes turbulent flow around the rocks.

Symbiosis [P]
A close biological interaction between species, often beneficial. The lichen is composed of bacteria or algae in a symbiotic relationship with a fungus.

Whitten effect [P]
Female mice can simultaneously enter oestrus (become ready to reproduce) as a result of pheromones in the urine of male mice.

The Coastline

The Coastline

Archie's law [L]
Relates the electrical conductivity of rock to how porous and saturated in brine it is. This law is used to estimate fossil fuel quantities when drilling offshore.

Archimedes' principle of flotation [L]
The force supporting a boat or other floating body is equal to the weight of the water the body displaces. This keeps the boat afloat.

Boyle's law [L]
The pressure of a gas increases as its volume decreases. As the foot pump is pressed down it increases gas pressure, inflating the boat.

Charles' law [L]
When heated, a gas expands. In the speedboat's engine, expanding hot gas pushes the pistons in the cylinders, powering the engine.

Conservation of momentum [L]
Momentum – the 'oomph' of movement – is conserved. When the bat hits the ball, momentum is transferred to the ball, sending it off in flight.

Dalton's law of partial pressures [L]
Air is a mix of different gasses – the total pressure is the sum of the pressures of each gas. The air the boy breathes combines partial pressures.

Fick's law of diffusion [L]
Molecules in the air travel very quickly, but undergo many collisions, slowing progress. Fick's law describes how smells travel through the atmosphere.

Newton's first law [L]
Something remains in motion at constant speed unless acted on by a force. The surfer keeps moving despite hitting the wave.

Green's law [L]
Describes mathematically how waves increase in height and come closer together as the water becomes shallower near the shore.

Newton's third law [L]
Every action has an equal and opposite reaction. As the boat's propeller pushes water backwards, this pushes the boat forward.

Henry's law [L]
Shows how the amount of gas dissolved in liquid changes with its partial pressure. Divers get decompression sickness (the bends) when gas dissolved in body fluids emerges as bubbles.

Stokes' law [L]
Describes the drag on a body moving through a fluid. The beach ball moves slowly as its large surface area produces significant drag, but its momentum is low.

Law of superposition [L]
The concept that lower layers of rocks formed before the layers that are above them. The higher strata seen on the cliff are younger than the lower ones.

Adiabatic cooling [P]
When the pressure suddenly decreases in an enclosed system, temperature drops. A mist forms on the fizzy drink can as it cools water vapour from the air.

Baywatch principle [P]
It's better to take a longer route if more of the route is quicker to traverse. The lifeguard first runs along the shore before entering water. This is why light bends when it goes, say, from air to glass.

Emulsification [P]
When two liquids that don't usually mix are successfully mingled by breaking one up into small droplets. Before freezing, the ice cream is an emulsion.

Bruun rule [P]
A formula to estimate the rate at which a shoreline recedes as a result of sea level rise.

Evaporative cooling [P]
When a liquid evaporates it takes energy from its surroundings, cooling them. The water on the swimmer's skin makes her feel cold after leaving the sea.

Diffraction [P]
Waves bend and spread out as they pass through a narrow opening. The waves expand into the harbour from the open sea.

Faunal succession [P]
The age of rock strata can be determined by the fossils found in them because they are deposited at time of death.

Dzhanibekov effect [P]
Describes the interaction of rotation around different axes. When the player flips her racket through 360 degrees, it also rotates to bring the other face up.

Gravity waves [P]
As wind blows across the sea, it moves the water, and waves form when gravity restores the water's position. These are gravity waves.

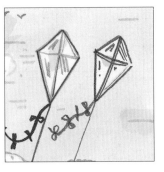

Heat capacities [P]
As sea has greater heat capacity than land, it warms more slowly. The cool sea air moves towards warmer, less dense air onshore, generating the sea breeze that flies the kites.

Longshore drift [P]
The movement of sand along the coast by wind-generated sideways currents. The shape of the beach has been changed by longshore drift.

Incandescence [P]
Giving off light due to heat. This increases the energy of electrons, which is then lost as photons of light. The fire is incandescent.

Magnus effect [P]
The ability of spin to bend the path of a ball by setting up a pressure differential. The spinning beach ball moves off its expected course.

Kelvin wake pattern [P]
When a bird or a boat without a propeller moves through calm water, it can produce a distinctive 'Kelvin wake' pattern.

Mandelbrot's coastline paradox [P]
A shorter measuring stick goes into more nooks and crannies, so a coastline has no definite length – the shorter the measure, the longer the coastline.

Kelvin–Helmholtz instability [P]
A turbulence effect generated when there is a velocity difference in fluid flows. The clouds make the instability visible in air that passed over the mountain.

Moulting [P]
An animal shedding an outer layer, often as the organism grows. The crab's shell cannot grow, so it moults and produces a larger shell.

Non-homogeneous mass dynamics [P]
The aerodynamics of the shuttlecock means that after being hit it rotates to be nose forwards, oscillates and then stabilises. This reflects its non-homogeneous makeup.

Plate tectonics [P]
The plates of the Earth's surface gradually move. As these plates collide, crumpling the Earth's crust, hills can be formed.

Orographic clouds [P]
As air is forced up the mountainside, it cools rapidly. This cooling causes water vapour to produce droplets, inducing cloud formation.

Plateau–Rayleigh instability [P]
The mechanism by which surface tension breaks a stream of slow-flowing liquid into droplets. At low pressure, the showerhead rains drops of water.

Pascal's principle [P]
A pressure change is transmitted through an incompressible fluid, so the large piston in the water pistol produces a high-speed spray from the much narrower jet.

Polarisation [P]
Sunlight reflecting off surfaces such as the sea becomes polarised. The Polaroid sunglasses reduce glare by cutting out some polarised light.

Piezo resistance [P]
A change in electrical resistance due to pressure. The diver's depth gauge uses a piezoresistive detector.

Principle of lateral continuity [P]
The observation that sedimentary layers were originally continuous, so rocks that were initially continuous but now have a gap have been eroded in the gap.

Rayleigh scattering [P]
Atoms scatter light by absorbing photons and re-emitting them in different directions. Air molecules scatter blue light more, producing the blue sky.

Thermal lag [P]
Seawater takes time to warm up, and cool down again, so is still chilly at midsummer, when sunshine is hottest, and still warm at the autumn equinox.

Refraction [P]
Waves change direction when they move into a region where they travel at a different speed (see Baywatch principle, page 94). As waves enter shallow water they are refracted.

Tidal forces [P]
The mechanism by which gravitational attraction causes a gradient. The Moon's tidal force causes high and low tides, making significant differences in sea level.

Reverse osmosis [P]
Forcing liquid through a semi-permeable membrane against osmotic pressure. This is used in the desalination plant to push water through, leaving salt behind.

Transverse wave [P]
A wave where the oscillation is at right angles to the direction of travel. Water waves are transverse waves.

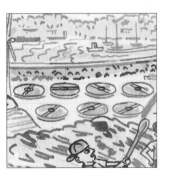

Stokes drift [P]
The rate at which an object is carried by fluid flow. Objects that are floating on the water are carried along slowly by the waves.

Venturi effect [P]
The effect is when speed increases and pressure drops as an incompressible fluid passes through a constriction. The scuba diver's regulator reduces pressure.

The Continent

The Continent

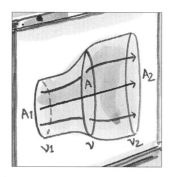

Betz's law [L]
The maximum power of a wind turbine is around 59.3% of the wind energy (because extracting all the energy would stop the air, preventing further movement).

Gibrat's law [L]
Predicts the growth of a city is independent of its size, but is distributed according to a log-normal distribution (where the logarithm of the value is distributed on a bell-shaped curve), though this is disputed.

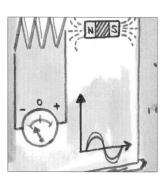

Buys Ballot's law [L]
In the northern hemisphere, with your back to the wind, low pressure is front and left, high pressure is front and right – the reverse applies in the south.

Adiabatic cooling [P]
The mechanism for the formation of cloud forms such as lenticular, when air expands without exchanging heat with its surroundings, dropping in temperature.

Faraday's law of induction [L]
The wind turbine's generator works by electrical induction – the law predicts the electrical energy produced depends on the rate at which the magnetic field changes.

Air mass [P]
Large volume of air with a relatively consistent temperature and water vapour content. The interaction of these masses governs our weather.

Asteroid impact crater [P]
Natural depression produced by a body impacting a planet or moon from space. Some, such as the 150-kilometre/93-mile-wide Chicxulub crater associated with the dinosaur extinction, are vast.

Chaotic weather system [P]
Weather systems are mathematically chaotic. Meteorologists run multiple 'ensemble' forecasts with slightly different starting conditions, as shown on the children's weather maps.

Buoyancy [P]
The upward force that keeps the boat afloat, where the weight of the boat is balanced out by the pressure difference of the column of water beneath it.

Combustion [P]
A chemical reaction giving off heat where a fuel reacts with oxygen. Natural combustion – for example, forest fires – is often started by lightning strikes.

Butterfly effect [P]
Suggests a small input, such as the flap of the butterfly's wings met in The Countryside (see page 83), can produce a significant change in the chaos-driven weather system.

Conductivity [P]
The capability of a substance to conduct electricity – the inverse of its resistivity. The power cables are designed to have high conductivity to minimise loss to heat.

Catenary [P]
The shape of a chain or cable suspended between two points under gravity. The powerlines form a catenary between pylons.

Depression of boiling point [P]
The lower the atmospheric pressure, the lower the boiling point. At 4,500 metres (14,760 feet), water boils at 84.5°C (184.1°F), making for an unsatisfactory cup of tea.

Electric discharge [P]
The ability of electricity to flow through air. A high voltage strips electrons from atoms, which enables them to carry a current such as a lightning bolt.

Exponential growth [P]
Growing at a rate proportional to the value growing, for example, doubling each time unit. The electrons in the lightning fall in a cascade, growing the charge exponentially.

Electromagnetic pulse [P]
A short duration burst of electromagnetic energy. The pulse caused by the lightning destroys nearby electronic devices, including the telephone.

Fluid dynamics [P]
The river's meanders (named after the River Meander in Turkey) are produced by the flow of the water, moving sediment from the outside to the inner bank.

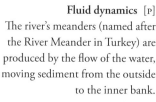

Electromagnetism [P]
The fundamental force responsible for the interaction of light and matter. Radio waves from the tower are electromagnetic waves.

Glaciation [P]
Deep, U-shaped, straight-sided valleys were produced by glaciers, scouring the earth as the ice slowly moved down a slope.

Erosion [P]
When flowing water or air gradually removes the surface of soil or rock. The cliff face is gradually retreating due to erosion.

Igneous rock formation [P]
Crystalline or glass-like rocks, formed from material melted by the Earth's inner heat to form magma, which then may be extruded as lava from volcanoes.

Ionisation [P]
The removal or addition of electrons from or to atoms to produce electrically charged ions. The lightning ionises the air, which makes it a conductor.

Lake-effect snow [P]
Cold wind passing over warmer water picks up water vapour and rises. When the water vapour meets cooler air, it produces snow.

Isostatic equilibrium [P]
Analogous to buoyancy, the balance between the gravitational pull on a mountain and the upthrust from the Earth's crust. Explains the heights of some mountains.

Lapse rate [P]
Describes how temperature falls with altitude (about 6.5°C per kilometre/ 18.8°F per mile), resulting in snow all year round on mountain tops.

Jet stream [P]
Fast-moving streams of air in the upper atmosphere. When the aircraft enters the jet stream its ground speed increases, making journeys faster and more fuel efficient.

Lightning strike [P]
When lightning results in an electrical discharge through a person, often producing severe injury or death. The most struck individual, park ranger Roy Sullivan, survived seven lightning strikes.

Katabatic wind [P]
Cold, high-density air that streams down a mountain as it falls into warmer, lower density air under the force of gravity.

Massenerhebung effect [P]
Mountains surrounded by higher peaks, which act as wind breaks, will have higher treelines than isolated mountains, as heat is held in.

Metamorphic rock formation [P]
Rock that is formed by a combination of heat and pressure, but without the rock melting.

Pingos [P]
Shallow, conical, ice-cored hills produced in an area with permafrost, due to the expansion of freezing water, either from underground lakes or aquafers.

Natural fission reactors [P]
If sufficient uranium is present underground it can undergo a chain reaction, producing considerable heat as a natural fission reactor, notably in Oklo, Gabon.

Plate tectonics [P]
The movement of plates on the surface of the Earth. When one plate slides over another plate it can rise up, causing mountain ranges, such as the Himalayas.

Oxidation [P]
Reaction of a substance with oxygen in the air. The oxidation of iron produces the rust on the iron bridge.

Potential difference [P]
The difference in electrical voltage between one location and another. The potential difference in voltage between the cloud and the ground triggers lightning.

Pareto principle [P]
In many cases, 80% of outcomes come from 20% of causes. In the town, 80% of the property is owned by 20% of the population.

Principle of intrusive relationships [P]
A geological method for the dating of rocks. When igneous rock cuts across sedimentary rock, the igneous is younger. Such intrusion produces formations like this batholith.

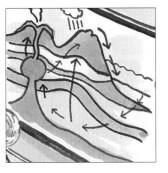

Sedimentary rock formation [P]
Rocks formed by the deposition and compacting of sediment, such as sand, followed by the particles being cemented together by chemicals in groundwater.

Temperature inversion [P]
When warm air is held above cooler air in the atmosphere. This inversion can trap pollution and forms low-lying clouds.

Skin effect [P]
An alternating current, such as in the power transmission cables, induces currents that push the main flow to the outside of the conductor, increasing resistance.

Triboelectric effect [P]
Static electricity generated by rubbing. Rubbing a balloon on the cat produces the same electrical effect as that which occurs between airborne ice crystals on a vast scale, producing lightning.

Solar energy [P]
Most of the Earth's energy, driving the weather systems, comes from the Sun. The solar farm in the model uses this electromagnetic energy.

Tsunami [P]
Waves on a body of water that are produced by a sudden shift in a mass of water, for example, due to an earthquake or volcanic eruption.

Subduction [P]
When one tectonic plate dives beneath another plate it can push up hot material, producing a chain of volcanoes.

Weather front [P]
The boundary between air masses at different temperature pressures, producing weather effects. Fronts are indicated by the semicircles and triangles on weather maps.

The Earth

The Earth

Newton's law of gravitation [L]
Newton's law tells us that gravitational attraction is towards the centre of gravity of a body. This applies wherever on the Earth a person is located.

Alternative projection map [P]
There are many ways to project a map of the spherical Earth on to a flat surface. In azimuthal projection, distances and directions from a central point are accurate.

Second law of thermodynamics [L]
Living things, like the girl, have less entropy (disorder) than a collection of elements, yet the law says entropy won't decrease. The energy from the Sun makes this possible.

Artificial satellites [P]
We make widespread use of artificial satellites in orbit around the Earth for communications, navigation, weather monitoring and more.

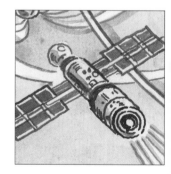

Acid rain [P]
Pollutants in the atmosphere, such as sulphur dioxide and nitrogen oxides, dissolve in rainwater, producing acids that can damage animals, plants and buildings.

Aurora effects [P]
Auroras occur when the solar wind – charged particles from the Sun – overcomes the protection of the Earth's magnetic field against incoming charged particles, generating light by stimulating atmospheric molecules.

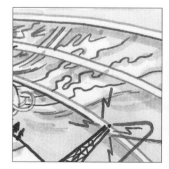

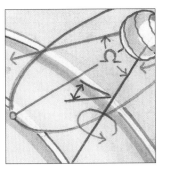

Barycentric rotation [P]
The Moon and the Earth orbit their joint centre of mass or 'barycentre'. For the Earth and Moon this is inside the Earth.

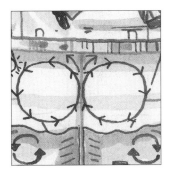

Convection currents [P]
Vast plates on the Earth's surface gradually move in the process known as plate tectonics. This is due to convection currents in the Earth, caused by temperature differences.

Carbon dioxide as greenhouse gas [P]
Carbon dioxide, produced by burning fossil fuels and volcanoes, is a major greenhouse gas, which re-emits infra-red radiation from the Earth back to the surface.

Coriolis force [P]
Force produced by rotation on the surface of a large rotating body, such as the Earth, at right angles to a moving object's direction of travel.

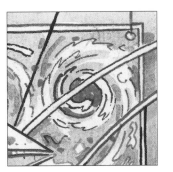

Chaos theory [P]
Describes systems where a small change in initial conditions produces large differences in outcome. The weather map shows a chaotic system.

Day and night [P]
The Earth's rotation over 24 hours means that the part lit by the Sun moves around the planet, creating day and night.

Climate change [P]
Alterations in the climate due to natural or artificial causes. Currently the climate is warming, causing sea level rise and changing habitats, often strongly affecting islands.

Discovering Earth's structure [P]
The Earth's structure is deduced by the deflection and absorption of shock waves passing through it from phenomena such as earthquakes. The seismograph picks up these vibrations.

Earthquakes [P]
Vibrations in the Earth's surface, usually caused by the interaction of moving plates of the Earth's crust.

Equatorial bulging [P]
As the Earth rotates, centrifugal force means that the crust is pushed outwards towards the equator, making Earth an oblate spheroid, rather than a sphere.

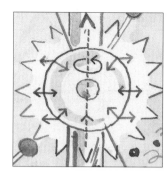

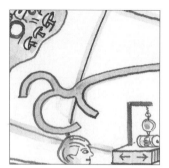

El Niño effect [P]
A change in the temperature distribution of the Pacific that switches water flows, radically altering weather on both sides of the ocean.

Equinoxes [P]
Literally 'equal nights', the two points in the Earth's orbit around the Sun when the centre of the Sun is directly over the Earth's equator.

Electromagnetism [P]
The fundamental force keeping atoms together. Electromagnetism prevents objects passing through each other, making it possible to stand on the walkway.

Fast carbon cycle [P]
The cycle of carbon absorbed from the atmosphere by growing plants, which are consumed by animals. The carbon is released back into the atmosphere as organisms decay.

Eötvös effect [P]
A change in gravitational force due to a centrifugal effect. The eastward-moving ship feels less gravitational pull than the westbound ship from the Earth's rotation.

Future sea levels map [P]
Climate change causes sea levels to rise, both because water expands and ice melts. The map shows possible future coastlines as water rises.

Geosynchronous orbit [P]
Orbiting the Earth at the same speed as the Earth's rotation. Satellites in geosynchronous orbit stay above the same points on the Earth's surface.

Ice ages [P]
Periods when global cooling saw polar ice extend over significant portions of the continents. The mammoth was locked into ice during an ice age.

Global warming [P]
As a result of climate change, global temperatures are increasing. This effect is strongest at the poles, where large amounts of ice are melting.

Magnetic poles [P]
The Earth has a magnetic field due to its spinning molten iron and nickel core. The magnetic poles are near but not on the planet's rotational poles.

Greenhouse effect [P]
Mechanism by which gas molecules such as carbon dioxide and methane reflect infra-red back to the Earth's surface, warming the planet.

Magnetosphere [P]
The Earth's magnetic field, shielding the planet from solar wind. Otherwise, this stream of charged particles would expose the surface to radiation and strip the atmosphere.

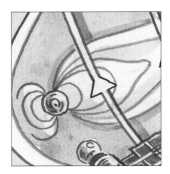

Heaviside layer [P]
A layer of electrically charged gas in the upper atmosphere that bounces radio waves around the Earth's curvature, reaching further than would otherwise be possible.

Methane as greenhouse gas [P]
Methane (natural gas) is a stronger greenhouse gas than carbon dioxide. A significant source is the digestive system of ruminants such as cows.

Near Earth asteroids [P]
Relics of the early solar system ranging in size from dust to kilometres across, with orbits around the Sun that come near to that of the Earth.

Ozone layer [P]
A layer of ozone (O_3) gas in the stratosphere (where the shuttle is passing through). The ozone layer absorbs incoming ultraviolet from the Sun, reducing otherwise dangerous levels on the Earth.

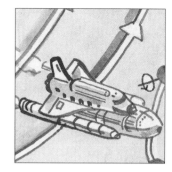

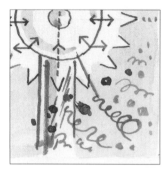

Neutrinos [P]
Particles emitted by the Sun that interact so weakly with matter that they pass right through the Earth and have been used to photograph the Sun at night.

Paleomagnetism [P]
Magnetic fields in iron-bearing rocks face in unexpected directions compared to the Earth's field. This makes it possible to trace old positions of tectonic plates.

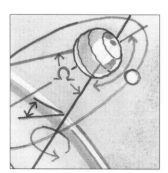

Nitrogen cycle [P]
Natural cycle where nitrogen is 'fixed' as nutrients from the atmosphere by bacteria in a symbiotic relationship with plants and later returned to the atmosphere by other bacteria.

Rock cycle [P]
Rocks transition through three main types – sedimentary, metamorphic and igneous – as they move through sections of the Earth at different temperatures and pressures.

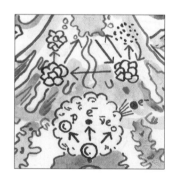

Obliquity [P]
The angle between a body's rotational axis and its path around an orbit. It is the Earth's obliquity that produces the seasons.

Sea floor spreading [P]
As tectonic plates shift, the sea floor can spread out, forming new oceanic crust as volcanic magma cools.

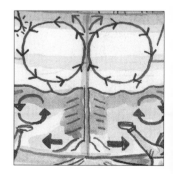

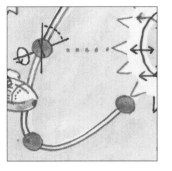

Slow carbon cycle [P]
In the slow carbon cycle, carbon in the ocean from shells and other organic matter is deposited, becomes part of rock and is eventually re-emitted by weathering and volcanic action.

Trade winds [P]
The prevailing east-to-west winds in the northern hemisphere that were the basis for using sail-powered trading ships.

Solstices [P]
Points in the Earth's orbit that put the Sun at its most northerly and southerly point in the sky.

Volcanoes [P]
Where pressure from tectonic activity forces molten rock (magma) up through the Earth's crust, resulting in it pouring out as lava.

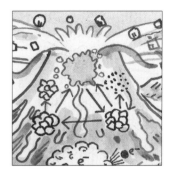

Strong interaction [P]
The force keeping atomic nuclei together and holding in place the quarks making up protons and neutrons. The diagrams show particle interactions involving the strong force.

Water cycle [P]
Natural cycle where the Sun vaporises water from the oceans. The water falls as rain and runs back to the ocean, often after use by biological organisms.

Thermohaline circulation [P]
Large-scale movements of warm water in the oceans that modify local weather patterns. Best known is probably the Gulf Stream.

Weak interaction [P]
Responsible for nuclear decay. The weak interaction is indirectly responsible for volcanoes, as the heat source for such activity is mostly the Earth's internal nuclear reactions.

The Solar System

$$P_y^2 = a_{AU}^3$$

The Solar System

Conservation of angular momentum [L]
Why everything in the universe spins. As material is pulled together by gravity, any rotating motion is amplified, like the spinning dancer pulling her arms in.

Kepler's second law [L]
A line joining a planet to the Sun sweeps past equal areas in equal times. The bungee cord keeping the birdcatcher in his orbit does this.

Dermott's law [L]
The orbital period of major moons is proportional to a constant raised to the power of the order (nearest to furthest) the moon is away from the planet.

Kepler's third law [L]
As shown on the music stand, the square of a planet's orbit period is proportional to the cube of the distance from the centre of the ellipse to a focus.

Kepler's first law [L]
A planet's orbit is an ellipse, with the Sun at one of the ellipse's foci (the equivalent of a circle's centre). The birdcatcher is acting as a planet around the singer's Sun.

Kirchoff's second law of spectroscopy [L]
A low-density gas, such as the surface of the Sun, emits a spectrum of colours, rather than a single colour.

Kirchoff's third law of spectroscopy [L]
Light composing a continuous spectrum, passing through a relatively cool, low-density gas, such as the Sun's atmosphere, will have black lines where elements absorb specific frequencies.

Black body radiation [P]
The distinctive distribution of light wavelengths given off by a heated body. Sunlight, produced by the Sun's hot surface, is approximately black body radiation.

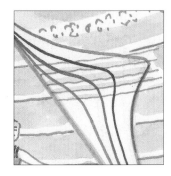

Accretion [P]
The process by which solar systems form as a mass of gas and dust, collectively spinning, is pulled together by the force of gravity.

Comets [P]
Bodies of dirty ice orbiting the Sun on long elliptical orbits. As comets near the Sun they warm and give off gasses, which can be seen as glowing tails.

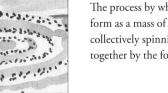

Albedo [P]
The degree to which an astronomical body reflects light. The Earth's albedo is increased by clouds and ice.

Eclipses [P]
When one astronomical body gets between the Sun and another body, casting a shadow. When eclipsed, the Moon is red as some light is scattered by the Earth's atmosphere.

Balmer spectral lines [P]
Dark lines in the spectrum of light from the Sun or from other stars, showing the presence of hydrogen, as each element absorbs light of specific colours.

Ecliptic [P]
The plane of the Earth's orbit around the Sun.

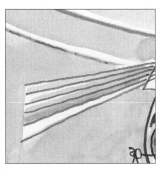

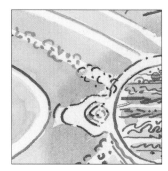

Electromagnetic radiation [P]
A stream of energy carried by an interaction between electricity and magnetism. Many astronomical bodies, including Jupiter, produce radio waves, part of the electromagnetic spectrum.

Greenhouse effect [P]
Venus should be like a hot Earth, but a runaway greenhouse effect from its CO_2-rich atmosphere produces temperatures of around 460°C (860°F).

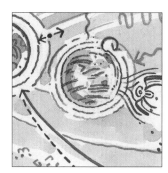

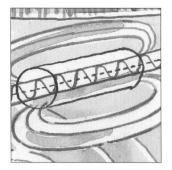

Faraday effect [P]
Each light particle (photon) has a direction at right angles to its movement, known as polarisation. Magnetic fields such as Jupiter's rotate sunlight's polarisation.

Heliosphere [P]
The region of influence of the Sun that is the extent of the solar wind. NASA's two Voyager probes have reached the edge of this region.

Giant impact hypothesis [P]
Our best idea for why we have an unusually large Moon is that a planet-sized body smashed into the young planet Earth, blasting out a large chunk.

Hill sphere [P]
The distance from the Earth where the Earth's gravitational field no longer dominates. To orbit the Earth, a satellite needs to be within the Hill sphere.

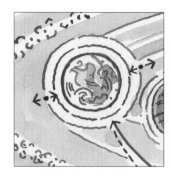

Gravity [P]
The planets' orbits are determined by the force of gravity. The existence of Neptune was predicted by its gravitational impact on other planets' orbits.

Kirkwood gaps [P]
Gaps in the asteroid belt between Mars and Jupiter where there is an 'orbital resonance' with Jupiter, meaning, for instance, they do three orbits of the Sun for one of Jupiter.

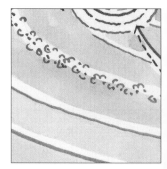

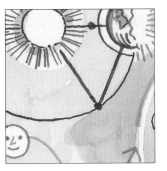

Lagrange points [P]
Points of gravitational stability from interaction of two massive bodies. Five such Earth/Sun points allow objects such as satellites to keep in place without drifting away.

Nice model [P]
A model for the development of the early solar system (devised in Nice) suggesting the giant planets were initially closer to the Sun and migrated outwards.

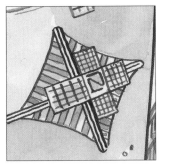

Light pressure [P]
Photons of light exert a tiny amount of pressure. In space, this means that a satellite can use solar sails as a means of propulsion.

Nuclear fusion [P]
The Sun is powered by nuclear fusion, where atomic nuclei join together to form heavier elements, releasing energy. The primary reaction is hydrogen fusing to produce helium.

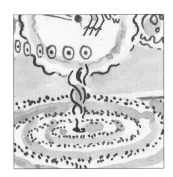

Martian meteorites [P]
When Mars is hit by an asteroid, it can blast out chunks of rock that eventually arrive at the Earth as Martian meteorites.

Oberth effect [P]
The so-called gravitational slingshot effect, where a space probe passes around a planet, picking up speed from the planet's orbital motion around the Sun.

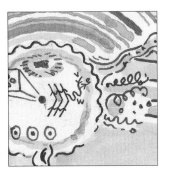

Neutrino oscillation [P]
The Sun's nuclear reactions produce electron neutrinos (see page 116), but far fewer than expected. This is because neutrinos change 'flavour' in flight to different kinds, a process called oscillation.

Oort cloud [P]
A region of icy bodies far outside the orbits of the planets. It is thought some comets with long orbital periods originated in the Oort cloud.

Orbital resonance [P]
Some of the moons of Saturn have orbital resonance, with orbital periods amounting to multiples of each other, a resonance effect a little like pushing someone on a swing.

Retrograde rotation [P]
Most planets rotate in the same direction because their rotation came from the same contracting material, but Venus spins the opposite way, hit by an asteroid long ago.

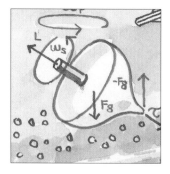

Precession [P]
When the direction of a rotating body's axis, or orbiting body's orbit, changes with time. This is seen in the gradual shift of planets' orbits – or this spinning top.

Self-organising system [P]
Jupiter's giant red spot, bigger than Earth, has lasted hundreds of years: it is an effect of chaotic systems producing long-lasting fluid flows, like the Gulf Stream on the Earth.

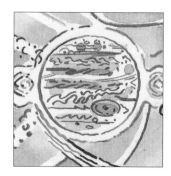

Quantum tunnelling [P]
Because the location of quantum particles is probabilistic, they can penetrate an apparently unbreachable barrier. Without this effect, hydrogen ions in the Sun can't get close enough to fuse.

Sidereal period [P]
The time for a body to complete an orbit relative to the stars. The Moon's sidereal period is different from its orbital period seen from the moving Earth.

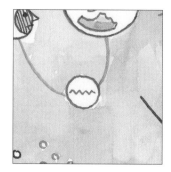

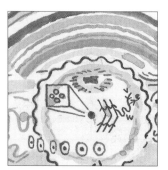

Rayleigh–Bénard convection [P]
A fluid heated from below can produce a convection pattern looking a little like frogspawn. This convection process happens on the Sun's surface.

Solar flares and coronal mass ejections [P]
Solar flares, sudden bright spots on the Sun, are often accompanied by coronal mass ejections, where charged material blasts from the Sun's surface.

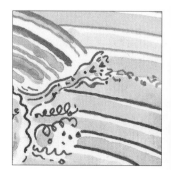

Solar wind [P]
A constant stream of electrically charged particles that flows outward from the Sun in all directions.

Wilson effect [P]
The eighteenth-century realisation that sunspots were on the Sun's surface, rather than orbiting, as they appear to flatten when seen at an angle near the poles.

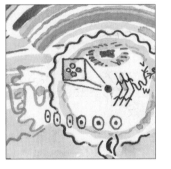

Strong interaction [P]
The force that holds together basic particles in the atomic nucleus. Some of the energy is released when nuclei fuse together, providing the Sun's power source.

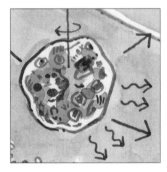

X-ray fluorescence [P]
Venus gives off X-rays because X-rays produced by the Sun are absorbed by its atmosphere and re-emitted, a process known as fluorescent scattering.

Sunspot cycles [P]
Sunspots, regions of the Sun's surface at lower temperature caused by magnetic effects, come and go, with maximum occurrence roughly every 11 years.

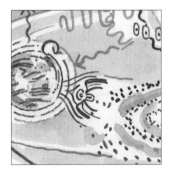

Yarkovsky–O'Keefe–Radzievskii–Paddack effect [P]
Known as YORP for short, the effect describes changes in the speed of rotation of small bodies such as asteroids due to absorbing and re-emitting solar radiation.

Tidal lock [P]
Because the Earth's gravity distorts the shape of the Moon, it has gradually changed the Moon's rotation period so that the Moon keeps the same face to Earth.

Zeeman effect [P]
A strong magnetic field can split a spectral line – light emitted by atoms at a particular frequency – into several different frequencies. Sunspots cause this effect.

The Entire Universe!

The Entire Universe!

Conservation of angular momentum [L]
The amount of rotational 'oomph' is conserved. This is why skaters spin faster when they pull in their arms and why galaxies form in spirals.

Inverse square law [L]
An effect that reduces with the square of the distance. A light's brightness drops off as an inverse square: this is used to measure the distance to stars and galaxies.

Coulomb's law [L]
Describes the force between electrically charged particles. Ion thrusters, used to shift satellite positions in orbit, use electrical repulsion to push out reaction mass.

Newton's third law [L]
Enables rockets to work. As the rocket's exhaust is pushed out of the spaceship, this produces an equal and opposite force forward on the spaceship.

Hubble's law [L]
Nearest neighbours apart, other galaxies are moving away from ours at a speed that grows with the distance. This shows that the universe is expanding.

Snell's law of refraction [L]
Describes how a beam of light changes direction as it moves between mediums where it has different speeds. The lenses of the projector focus due to Snell's law.

Uncertainty principle [L]
The uncertainty principle says that particles briefly appear and disappear in space – like the whack-an-alien game, they disappear so quickly they're hard to spot.

Chandrasekhar limit [P]
The maximum mass that a white dwarf star can have and stay stable. Such stars with bigger masses will collapse to become neutron stars or black holes.

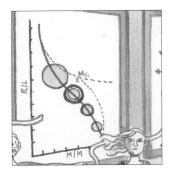

Astrometry [P]
Making measurements of positions in space. Astrometry is used to detect exoplanets – planets around other stars – by the wobble the planets' gravity causes on the stars' motion.

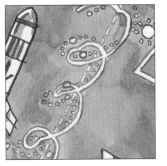

Cosmic microwave background radiation [P]
Light that began crossing the universe when it was 300,000 years old and first became transparent. This fills all space with faint microwave radiation.

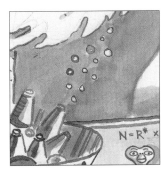

Black body radiation [P]
The electromagnetic radiation due to temperature from a non-reflective body. The cosmic microwave radiation that fills space is black body radiation with a temperature of 2.7 K (−270°C/−455°F).

Dark energy [P]
The expansion of the universe is accelerating. This requires energy to sustain it, known as dark energy. There are many theories, but the source is unknown.

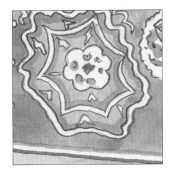

Black hole formation [P]
Black holes form when a large dying star collapses because it is too massive for internal pressure to resist the gravitational pull of its matter.

Dark matter [P]
Hypothetical undetected extra matter in the universe that does not interact electromagnetically. Deduced because galaxies rotate too quickly to hold together with their visible matter alone.

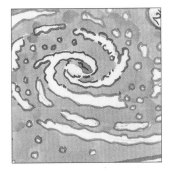

Doppler spectroscopy [P]
Mechanism to detect planets around other stars as the gravitational influence of the planet causes the star to undergo a Doppler shift, changing its colour. The little girl acts as a planet to her sister's star.

Exoplanet direct imaging [P]
Telescopes are becoming sufficiently powerful that some planets around stars other than the Sun can be seen directly, rather than detected by their effects. The ring is the composite path of the planet.

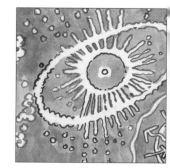

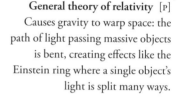

Drake equation [P]
A formula containing many unknowns that was devised to give an approximation to the number of potential alien civilisations in the Milky Way galaxy.

General theory of relativity [P]
Causes gravity to warp space: the path of light passing massive objects is bent, creating effects like the Einstein ring where a single object's light is split many ways.

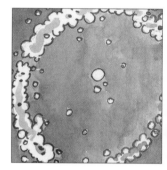

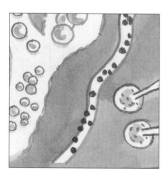

Eclipsing binaries [P]
One of the reasons stars can vary in brightness, where two stars orbit each other (binary stars), one passing in front of the other.

Gravitational waves [P]
Waves in the fabric of space-time, caused by massive events such as the spiralling collision of black holes, detected by tiny movements of devices on Earth.

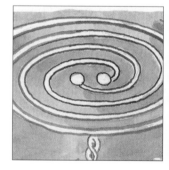

Eddington valve mechanism [P]
Also called Kappa mechanism, movement of layers in a star causing it to vary in brightness. Such variable stars are used as 'standard candles' to measure distances.

Hawking radiation [P]
Although a black hole's gravitational field is so strong even light cannot escape, it does give off faint Hawking radiation from particles produced by quantum effects.

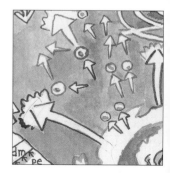

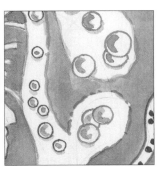

Hertzsprung–Russell diagram [P]
This diagrams shows the relationship of different star types and how stars evolve over time.

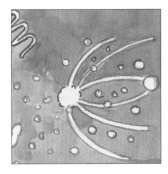

MOND [P]
Modified Newtonian Dynamics – an alternative explanation for the effect attributed to dark matter, where gravitational attraction is modified for very large bodies.

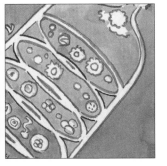

Inflationary theory [P]
Suggests the universe went through a brief period of extremely fast expansion early in its life. Helps explain its current appearance, but as yet without supporting evidence.

Multiverse theory [P]
The hypothesis that there have been multiple big bangs causing many universes to expand, like bubbles on the surface of a liquid.

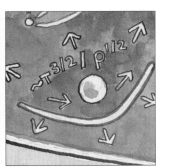

Jeans mass [P]
The mass at which a cloud of gas has sufficient gravitational attraction to contract and form a star.

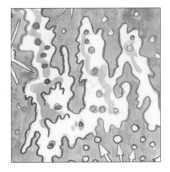

Nebula [P]
Fuzzy patches in space (from the Latin for 'cloud') – clouds of glowing gas, either where stars are forming or the debris of stellar explosions.

Lense–Thirring effect [P]
Effect of general relativity where a massive rotating body pulls space-time around with it like a rotating spoon in honey, a process known as frame dragging.

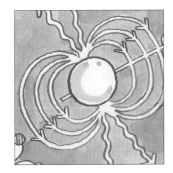

Neutron star formation [P]
The remains of a star that has undergone a supernova explosion, leaving a core so dense that a teaspoon-full has a mass of 100 million tonnes.

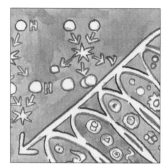

Nucleosynthesis [P]
The process by which stars and supernovas fuse hydrogen to produce helium, going on to make heavier substances, producing all the 94 natural elements.

Principle of homogeneity [P]
An assumption that on a large scale the universe is homogeneous, providing the same observational evidence wherever the observer is located.

Olbers' paradox [P]
If the universe were infinite we would expect stars in every direction. Edgar Allen Poe pointed out that finite light speed and lifetime of the universe means that we only see relatively nearby stars.

Principle of isotropy [P]
An assumption that on a large scale the universe is isotropic – the same observational evidence is available (and physical laws apply) in whatever direction you look.

Parallax [P]
We measure distances to closer stars by their apparent movement when seen from two viewpoints – just as objects shift when we cover one eye, then the other.

Pulsar [P]
A fast-rotating neutron star giving off regular radio pulses – when first observed, it was speculated these could be an alien transmission.

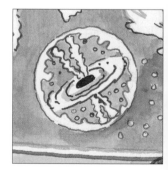

Plasma weapon [P]
One of the more feasible ray guns from science fiction, a plasma weapon would propel a beam or pulse of ultra-hot ionised gas by electromagnetic repulsion.

Quasar [P]
Short for 'quasi-stellar object': a supermassive black hole at the heart of a young galaxy, absorbing matter and emitting thousands of times the light of a normal galaxy.

Relativity of simultaneity [P]
The simultaneity of events in space depends on relative motion. The fast-moving skater observes one simultaneously dropped tray slightly earlier than the other.

Stellar core collapse [P]
A second mechanism to trigger a supernova, where the core of an extremely massive star collapses under the force of gravity, blowing off the rest of the material.

Rocket equation [P]
The equation describing the change in velocity possible from a rocket, dependent on the amount of reaction mass it carries: why rockets from Earth need multiple stages.

Tolman–Oppenheimer–Volkoff limit [P]
Collapsing stars producing a supernova can end up as a neutron star or a black hole – this is the maximum mass to form a neutron star.

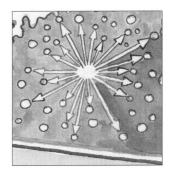

Runaway nuclear fusion [P]
One mechanism for stars to go supernova, in which a white dwarf star takes on extra matter and undergoes a sudden and drastic nuclear fusion reaction.

Transit photometry [P]
An alternative means of detecting planets around other stars where the star's light dims as the planet passes in front of it.

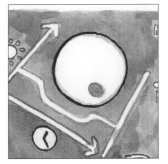

Sachs–Wolfe effect [P]
Variations caused in the cosmic microwave background radiation by gravitational redshift, where photons lose energy as they travel though a gravitational field.

Where the big bang happened [P]
Because every point in space emerged from the big bang, the location of the big bang is everywhere, including this location in the planetarium.

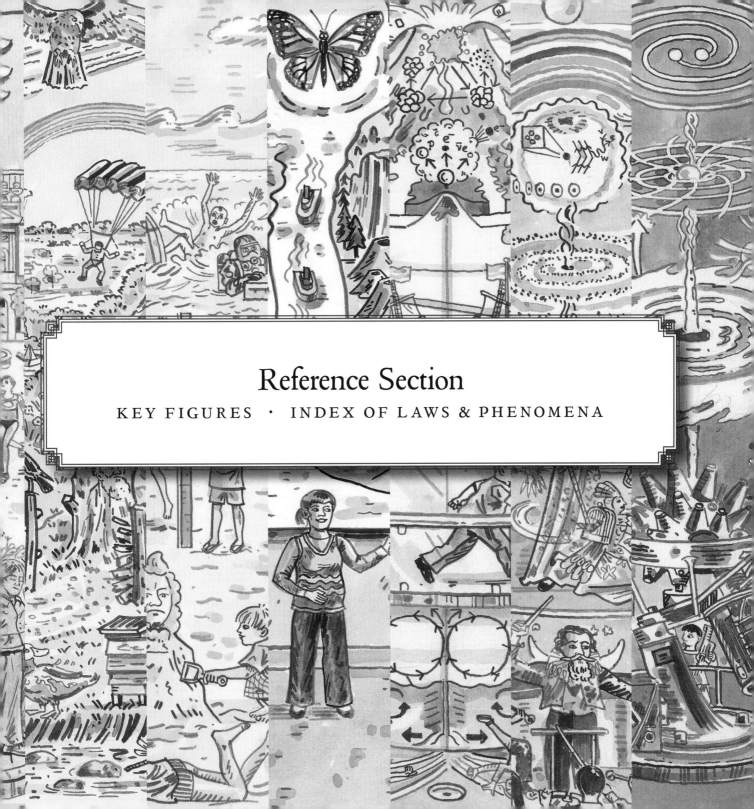

Reference Section

KEY FIGURES · INDEX OF LAWS & PHENOMENA

Key Figures

1

William Thomson (Lord Kelvin)
1824–1907

KEY DISCOVERY
Laws of thermodynamics

Born in Belfast on 26 June 1824, William Thomson, who would become the 1st Baron Kelvin, was highly successful both as a physicist and an engineer. After education at Cambridge he became Professor of Natural Philosophy at the University of Glasgow at the remarkably young age of 22. His greatest achievements were in the field of thermodynamics, but he also had a significant involvement in the project to lay a transatlantic telegraph cable, for which he was knighted in 1866. He came up with the concept of absolute zero, the lowest possible temperature, and would formulate an early version of the hugely important second law of thermodynamics. Thomson also contributed to arguments on the age of the Earth, considering that the planet must have been in existence for millions of years. He died at Largs on 17 December 1907.

2

Marie Curie

1867–1934

KEY DISCOVERY

Radiation

Born Maria Salomea Skłodowska on 7 November
1867 in Warsaw, Marie Curie studied physics at the
Sorbonne in Paris. Lacking postgraduate lab space,
she shared a lab with her soon-to-be husband,
Pierre Curie, marrying in July 1895. Two years later,
Curie began work on radioactivity. In 1898, she
discovered that the mineral pitchblende was a
powerful radioactive source. Working with Pierre,
she produced a substance 400 times as active as
uranium, later named polonium. Later that year they
identified another radioactive element, which they
called radium. The Curies won the 1903 Nobel Prize
in Physics for work on radioactivity, while Marie also
won the 1911 Chemistry Prize for discovering radium
and polonium. In 1906, Pierre was killed in a road
accident. Curie continued radioactivity work, and
developed mobile First World War X-ray units, later
pioneering medical applications of radioactivity. She
died on 4 July 1934 from over-exposure to X-rays.

3

Leonardo of Pisa (Fibonacci)

*c.*1170–*c.*1240–50

KEY DISCOVERY

Fibonacci series

Born in Pisa around 1170, Leonardo is better
known by his nickname Fibonacci from *filius
Bonaccii* – 'son of Bonaccio'. Fibonacci's father took
his son on trips to North Africa, where Fibonacci
was exposed to Indian numerals, used by Arab
mathematicians. In his 1202 book *Liber Abaci*,
Fibonacci introduced these numerals to the
West along with the zero. In that same book,
Fibonacci studied population increase. In his simple
model, rabbits take one month to mature, and an
adult pair produces a new pair (one male, one
female) every month. No rabbits die. Month by
month, the number of pairs goes 1, 1, 2, 3, 5, 8, 13 …
Each month's value adds together the two
previous values. Although not realistic for
population modelling, the series is reflected in
natural phenomena, such as seed distribution in
flower heads. After several decades working in
accounting and mathematical instruction,
Fibonacci died between 1240 and 1250.

4
Richard Feynman
1918–88

KEY DISCOVERY
Quantum electrodynamics (QED)

Born in New York on 11 May 1918, Richard Feynman was the physicist's physicist – a legend in his field, as much for his charismatic ability to communicate physics as his scientific genius. After graduating from the Massachusetts Institute of Technology (MIT), Feynman joined the Manhattan Project, the Second World War project to develop a nuclear bomb. Feynman's biggest contribution to physics was quantum electrodynamics (QED), the science of light and matter, for which he shared the Nobel Prize. The Feynman diagram, a tool he developed to help work on QED, proved essential in the development of the field. Feynman became well known for two publications. The notes from his undergraduate lectures, *The Feynman Lectures on Physics*, known as the red books, became surprise bestsellers for textbooks, while a collection of his memoirs in book form, *Surely You're Joking, Mr Feynman!*, had a different but equal appeal. Feynman died on 15 February 1988.

5
Lynn Margulis
1938–2011

KEY DISCOVERY
Endosymbiosis

Born in Chicago on 5 March 1938, Lynn Margulis developed her breakthrough idea, endosymbiosis, two years after receiving her PhD at the age of 29. Her theory was that small subunits of biological cells called mitochondria were originally bacteria, absorbed into another cell in a symbiotic relationship. Similarly, Margulis suggested that the chloroplasts enabling photosynthesis in many plants were once independent organisms. Initially her theory was rejected as it went against the incomplete Darwinian evolutionary ideas of the time. It was a decade before there was sufficient experimental evidence to make her ideas mainstream. Margulis also worked with James Lovelock on the Gaia hypothesis, suggesting that the Earth is a self-regulating system not unlike a huge organism. She continued to look for symbiotic relationships in the structures of cells and supported an unproven theory that larvae and adults of metamorphosing species evolved from different ancestors. Margulis died on 22 November 2011.

6
Isaac Newton
1642–1726

KEY DISCOVERIES
Laws of motion and gravity

Born at Woolsthorpe Manor in Lincolnshire on 25 December 1642 (old style, Julian Calendar), Isaac Newton proved exceptional when he attended Cambridge University. Soon after graduation in 1665, plague forced him to spend two years at home, when he claimed to have been inspired to think about gravity by seeing a falling apple. His first achievements were in optics, building an early reflecting telescope that brought him to the attention of the Royal Society, and explaining how white light is made up of the colours of the rainbow. In 1687, his masterpiece *Philosophiae Naturalis Principia Mathematica* was published, introducing his laws of motion and universal gravitation. It also made use of his new mathematical technique, the method of fluxions, now better known as calculus. Newton went on to success as Master of the Royal Mint. Knighted in 1705 for his political work, he died on 20 March 1726 (old style) in Kensington, London.

7
Michael Faraday
1791–1867

KEY DISCOVERIES
Electric motor, induction

Born into a poor family in London on 22 September 1791, Michael Faraday was apprenticed to a bookbinder at the age of 14. Self-educated from books and attending public lectures, Faraday got a job at the Royal Institution in London as Chemical Assistant in 1813. Proving a greater experimenter, he was promoted to Director of the Laboratory and became Professor of Chemistry in 1833. Despite his considerable achievements in chemistry, his greatest discoveries were in physics. In 1821 he discovered the phenomenon behind the electrical motor, while in 1831 he discovered electromagnetic induction, providing the mechanism for the electrical generator. He also came up with the concept of fields, which became central to physics theory and helped unite the concepts of electricity and magnetism. Faraday was also a consummate speaker and expanded the range of the Institution's popular lectures. He died on 25 August 1867 at Hampton Court near London.

8
Charles Darwin
1809–82

KEY DISCOVERY
Evolution

Born in Shrewsbury on 12 February 1809, Charles Darwin was educated at Cambridge with the intention of becoming a clergyman, but was fascinated by geology and botany. As a result, he was invited to join the five-year voyage of the HMS *Beagle* on a mission to map the coastline of South America. The trip, taking Darwin to South America and Australia, enabled him to collect a vast range of specimens. As a result of observations of variation in birds and tortoises on the Galápagos Islands, he speculated on the way in which species could develop. It was only 22 years later when Darwin received a letter from naturalist Alfred Russel Wallace, outlining a theory of natural selection, that he was pushed into publishing his famous book on evolution *On the Origin of Species* in 1859. After success with this and its sequel *The Descent of Man* in 1871, Darwin died on 19 April 1882.

9
George Stokes
1819–1903

KEY DISCOVERY
Stokes' law, Stokes drift

Born in Skreen, County Sligo on 13 August 1819, George Stokes was an eminent Victorian physicist, holding the post of Lucasian Professor at Cambridge (previously held by Isaac Newton and subsequently by Stephen Hawking) for 54 years. Educated at Cambridge, he spent his entire career at the university. He made significant advances in the understanding of polarised light and fluorescence (which he named) and did useful work on the variation of gravity across the Earth's surface, but his name is most closely associated with fluid dynamics, the science of the flow of liquids and gasses. The key equations in fluid dynamics, the equivalent of Newton's second law of motion, are the Navier–Stokes equations. These were correctly formulated by French engineer Claude-Louis Navier, but without good evidence. It was Stokes who, in 1845, put the equations on a scientific footing. Stokes was knighted in 1889 and died on 1 February 1903.

10
Alfred Wegener
1880–1930

KEY DISCOVERY
Continental drift/plate tectonics

Born in Berlin on 1 November 1880, Alfred Wegener faced an uphill battle to get his contentious theory accepted. Educated at Berlin, Heidelberg and Innsbruck, Wegener worked in meteorology and took part in four expeditions to Greenland. Early in his career, Wegener noticed that the shape of the east coast of the Americas was remarkably similar to the coasts of Africa and western Europe, fitting together like a giant jigsaw puzzle. He proposed that the continents had at one time been joined and had drifted apart. This was supported by similarities in rock types and fossil finds in the regions that would have been in contact. Wegener published his idea in 1915, but the concept was not accepted until a good 20 years after his death as it seemed so implausible. The mechanism, known as plate tectonics, is now entirely accepted. Wegener died when supplies ran out on his last Greenland expedition in November 1930.

11
Edward Lorenz
1917–2008

KEY DISCOVERY
Chaos theory

Born in West Hartford, Connecticut on 23 May 1917, Edward Lorenz was a mathematician and meteorologist who discovered the principle of chaos theory. After studying at Dartmouth College and Harvard, Lorenz went on to the Massachusetts Institute of Technology (MIT). It was there in 1961 that Lorenz made a remarkable discovery. He had been running a weather model on an early computer. This was stopped part way through – rather than restart the slow process, Lorenz re-input values taken part way through the run. To his surprise, the forecast developed in a totally different way to the first run. It turned out the system printed out fewer decimal places than it used in its calculations. The tiny difference led to a huge change in outcome – the essence of chaos theory. The title of a paper by Lorenz also gave us the term 'butterfly effect' for such an outcome. Lorenz died on 16 April 2008.

12
Johannes Kepler
1571–1630

KEY DISCOVERY
Laws of planetary motion

Born in Württemberg, Germany on 27 December 1571, Johannes Kepler worked on optics and logarithms, but is remembered for his laws of planetary motion. Educated at the University of Tübingen, he learned about the revolutionary theory of Copernicus, which put the Sun at the centre of the solar system. One of Kepler's early breakthroughs was to decide that the Moon was not a normal planet, describing it, with a term he coined, as a *satellite* of the Earth. Some of his cosmological theories seem a little bizarre now – he suggested that the orbits of the planets were determined by the size of regular solids such as a sphere, a tetrahedron and a cube fitted inside each other. However, his lasting legacy was his laws that state that planets move in ellipses with the Sun at one focus and describe the speed at which planets orbit. Kepler died in Regensburg on 15 November 1630.

13
Albert Einstein
1879–1955

KEY DISCOVERIES
Relativity, black holes, gravitational waves

The world's most famous scientist, Albert Einstein, was born in Ulm, Germany on 14 March 1879. Einstein had an early interest in science, but rebelled against formal schooling: within a year of his family moving to Italy, the 16-year-old left school, renounced his German citizenship and moved to Switzerland. After studying at Zurich Polytechnic he was unable to get a postgraduate post, taking a job at the Swiss Patent Office. While there in 1905, he published a series of papers establishing the size of molecules, introducing special relativity, explaining the photoelectric effect (laying foundations of quantum theory and winning him the Nobel Prize) and showing $E=mc^2$. His career peaked with his 1915 general theory of relativity, providing a new take on gravity. He went on to identify the mechanism of the laser and predict gravitational waves. Of Jewish heritage, Einstein left an increasingly hostile Germany in 1933 for the United States, dying on 18 April 1955.

Index of Laws & Phenomena

Author Biographies

ILLUSTRATOR

Adam Dant is an internationally renowned artist whose large-scale narrative drawings and prints can be found in the collections of the Victoria & Albert Museum, Tate, the Museum of Modern Art (MoMA) and the Musée d'Art Contemporain de Lyon as well as many leading private collections, including that of HRH The Prince of Wales. He was the recipient of a prestigious Rome Scholarship in etching and engraving, is a winner of the Jerwood Drawing Prize and in 2015 was appointed by the UK parliament as Official Artist of the General Election. His 2017 book *Maps of London & Beyond* was awarded the gold award at the 2018 International Creative Media Awards and first prize in the travel category at the 2019 *Catholic Herald* book awards.

AUTHOR

Brian Clegg is a Fellow of the Royal Society of the Arts, Member of the Institute of Physics, a regular broadcaster and lecturer, and author of over 40 bestselling science books, two of which have been listed for the Royal Society Prize for Science Book of the Year. He has worked in operational research and trained businesses in creativity and the development of ideas, and has written for numerous magazines and newspapers, ranging from *The Wall Street Journal* to *Physics World*. Recent books include *10 Short Lessons in Time Travel* (Michael O'Mara, 2021), *What Do You Think You Are?* (Icon Books, 2020) and *Dark Matter & Dark Energy* (Icon Books, 2019).

Created and produced by
Iqon Editions Ltd

And first published in the English language
in 2021 by
Ivy Press
An imprint of The Quarto Group
The Old Brewery, 6 Blundell Street, London N7 9BH, United Kingdom
T (0)20 7700 6700
www.quartoknows.com

British Library Cataloguing-in-Publication Data
A catalogue record for this book is available from the British Library.

ISBN: 978-0-7112-5678-1
E-book ISBN: 978-0-7112-5679-8

Publisher, concept and direction: David Breuer
Design and typography: Isambard Thomas
Project editor: Caroline Earle

Printed in China
10 9 8 7 6 5 4 3 2 1